EVOLUTION AND SYSTEMATICS OF THE ATLANTIC TREE RATS, GENUS *PHYLLOMYS* (RODENTIA, ECHIMYIDAE), WITH DESCRIPTION OF TWO NEW SPECIES

Golden Atlantic tree rat, *Phyllomys blainvilii* (photograph by L. P. Costa)

Evolution and Systematics of the Atlantic Tree Rats, Genus *Phyllomys* (Rodentia, Echimyidae), with Description of Two New Species

Yuri L. R. Leite

A Contribution from the Museum of Vertebrate Zoology, University of California at Berkeley

UNIVERSITY OF CALIFORNIA PRESS
Berkeley • Los Angeles • London

UNIVERSITY OF CALIFORNIA PUBLICATIONS IN ZOOLOGY

Editorial Board: Peter Moyle, James L. Patton, Donald C. Potts, David S. Woodruff

Volume 132

UNIVERSITY OF CALIFORNIA PRESS
BERKELEY AND LOS ANGELES, CALIFORNIA

UNIVERSITY OF CALIFORNIA PRESS, LTD.
LONDON, ENGLAND

Library of Congress Cataloging-in-Publication Data available upon request.

ISBN 0-520-09849-8

The paper in this publication meets the minimum requirements of
ANSI/NISO Z39.48-1992 (R 1997) (*Permanence of Paper*) {∞}

Para Yára Maria, Leonora e Sarah

Contents

Figures

Tables

Acknowledgments

This work is part of my dissertation submitted to the Department of Integrative Biology of the University of California, Berkeley, in partial fulfillment of the Doctor of Philosophy degree. It could not have been accomplished without vital help from several people. I thank J. L. Patton, B. D. Mishler, R. Byrne, L. H. Emmons, and P. Myers for providing critical comments that improved the quality of this monograph. I would also like to thank the Brazilian Conselho Nacional de Desenvolvimento Científico e Tecnológico (CNPq), Brasília, for granting me a predoctoral fellowship and partial support for fieldwork. Collecting permits were provided by the Instituto Brasileiro do Meio Ambiente e dos Recursos Naturais Renováveis (IBAMA), Brasília, and the Instituto Estadual de Florestas de Minas Gerais (IEF–MG), Belo Horizonte.

Although I did not know it at the time, this project started more than ten years ago when J. R. Stallings, E. L. Sábato, P. P. de Oliveira, M. C. M. Kierulff, and L. F. B. M. Silva introduced my wife, L. P. Costa, and me to the world of small mammals. Some specimens reported here were collected during a mammal survey of the Atlantic forest (Projeto Inventário Faunístico da Mata Atlântica), carried out in 1991–1992. G. A. B. da Fonseca headed this survey through Fundação Biodiversitas, Belo Horizonte, with funds from the John D. and Catherine T. MacArthur Foundation. L. P. Costa, M. T. da Fonseca, and I conducted field trips in the states of Minas Gerais and Espírito Santo, with help from M. A. Sábato, G. M. Moreira, and J. A. S. Silva. L. P. Costa and I carried out additional fieldwork in 1996 and 1998, with grants from the National Geographic Society, World Wildlife Fund, Brazil (to L. P. Costa), Museum of Vertebrate Zoology (MVZ), and the Berkeley Chapter of Sigma Xi. Alexandre M. Fernandes, D. C. Bianchini, L. G. Vieira, R. L. Dias, F. P. Santos, M. A. Sábato, R. T. Moura, J. L. Patton, and C. Patton provided crucial aid in the field. Although I collected several of the specimens in collaboration with L. P. Costa, I owe special thanks to scientists who allowed me to use specimens they obtained: H. G. Bergallo, R. Cerqueira, A. Christoff, V. Fagundes, M. T. da Fonseca, L. Geise, D. Huchon, M. Lara, M. A. Mustrangi, A. Paglia, R. Pardini, J. Pagnozzi, M. Passamani, J. L. Patton, R. Ribón, E. L. Sábato, M. A. Sábato, M. N. F. da Silva, J. C. Voltolini, and Y. Yonenaga-Yassuda. Despite our shared research interests, most of them provided free access not only to specimens (most still uncataloged), tissue samples, and DNA extracts but also to unpublished data and field notes. Their unselfish efforts are much appreciated and should serve as an example for the scientific community.

I would further like to thank the following curators for their hospitality and/or for making specimens under their care available for examination: S. L. Mendes (MBML); J. Ferigolo (MCNFZB); M. Miretzki (MNHCI); L. Flamarion, J. A. de Oliveira, and L. O. Salles (MNRJ); J. L. Patton (MVZ); L. C. M. Pereira (MZPUCPR); M. de Vivo (MZUSP); G. A. B. da Fonseca and A. B. Rylands (UFMG); A. Langguth (UFPB); G. M. del Giúdice (UFV); and M. D. Carleton (USNM). Alfredo Langguth, J. Clary, D. Kock, and S. Traenker kindly provided pictures of type specimens housed in London, Paris, Lyon, and Berlin. Jim Patton not only took pictures and measured every *Phyllomys* skull in the British Museum but also drew their toothrows and took better notes than I would have taken myself. L. H. Emmons provided crucial data on type specimens in European museums and shared photographs, notes, measurements, and her vast knowledge of tree rats. Given our similar research interests, we started a collaboration that resulted in a joint paper, and this report would have been incomplete without the information she obtained. She deserves special mention for her continuous support and encouragement throughout this work.

Special thanks to M. F. Smith for providing essential training and guidance in the Evolutionary Genetics Lab and R. E. Jones for preparing excellent specimens. Maria José de J. Silva helped in the preparation of chromosome slides, Karen Klitz prepared some of the figures, and W. Pang Chan from the Scientific Visualization Center gave continuous advice on scientific illustration. Many thanks to H. G. Greene, B. D. Mishler, W. A. Clemens, E. A. Lacey, M. F. Smith, E. Lessa, M. N. F. da Silva, A. Ditchfield, L. F. Garcia, M. D. Matocq, T. Hambuch, D. Meyer, M. M. Soares, M. Mahoney, J. Mobley, A. Chubb, R. Mueller, J. Rodríguez, and G. Parra for productive discussions, good advice, and support throughout this study. Very special thanks to J. L. Patton and L. P. Costa for their help and patience and for sharing with me their vast knowledge of and enthusiasm for small mammals.

Abstract

The Atlantic tree rats, genus *Phyllomys*, are arboreal echimyids found in eastern Brazil. Species of *Phyllomys* are of conservation interest because they are poorly known, have restricted geographic ranges, and are endemic to the Atlantic forest, one of the most threatened ecosystems in the world. Here I examine the diversity of the genus *Phyllomys* using genetic and morphological data. The goals are to elucidate the taxonomy of the group, infer phylogenetic relationships among species, and understand the processes that led to the present diversity and distribution. The Atlantic tree rats have been referred to as *Nelomys* or grouped with the Amazonian genus *Echimys*. Given that *Nelomys* is a junior synonym of *Echimys*, the name *Phyllomys* should be used. *Phyllomys* is monophyletic and readily diagnosable by unique dental characters, and there is no reason to include it with *Echimys*. There are currently ten described and three undescribed species of *Phyllomys*, two of which I describe here. Phylogenetic analyses of cytochrome b data corroborate the monophyly of the genus and indicate two geographically bounded clades: southern (*P. dasythrix* and *P.* aff. *dasythrix*) and northeastern (*P. brasiliensis, P. lamarum,* and *P. blainvilii*). The remaining species included in the analyses (*P. pattoni, P. nigrispinus,* and the two new species described here) are of uncertain placement. I address the tectonic, climatic, and vegetation changes in eastern Brazil since the Pliocene that set the stage for the diversification of *Phyllomys* and likely played a role in the phylogeographic patterns observed today.

Resumo

Os ratos de espinho do gênero *Phyllomys* são equimídeos arborícolas encontrados no leste do Brasil. As espécies de *Phyllomys* são de interesse especial para conservação, pois são pouco conhecidas, possuem distribuição geográfica restrita e são endêmicas à Mata Atlântica, um dos ecossitemas mais ameaçados do planeta. No presente trabalho, eu examinei a diversidade do gênero *Phyllomys* utilizando dados genéticos e morfológicos. Os objetivos foram elucidar a taxonomia do grupo, inferir relações filogenéticas entre as espécies e entender os processos que levaram à diversidade e distribuição atuais. Os ratos de espinho arborícolas da Mata Atlântica vêem sendo tratados como *Nelomys* ou aglutinados ao gênero amazônico *Echimys*. Dado que *Nelomys* é um sinônimo júnior de *Echimys*, o nome *Phyllomys* deve ser utilizado. O gênero *Phyllomys* é monofilético e facilmente diagnosticável por características dentárias únicas, não havendo nenhuma razão para sua inclusão em *Echimys*. Existem dez espécies descritas de *Phyllomys* e três não descritas, sendo que duas dessas foram descritas no presente trabalho. Análises filogenéticas de dados do citocromo b corroboraram a monofilia do gênero, e indicaram dois clados organizados geograficamente: sul (*P. dasythrix* e *P. aff. dasythrix*) e nordeste (*P. brasiliensis, P. lamarum* e *P. blainvilii*). As espécies restantes incluídas na análise (*P. pattoni, P. nigrispinus* e as duas novas espécies descritas aqui) apresentaram posições filogenéticas incertas. Mudanças tectônicas, climáticas e vegetacionais no leste do Brasil desde o Plioceno formaram o cenário da diversificação de *Phyllomys* tendo, provavelmente, desempenhado um papel importante nos padrões filogeográficos observados atualmente.

INTRODUCTION

THE GENUS *PHYLLOMYS*

The Atlantic tree rats of the genus *Phyllomys* are small- to medium-sized arboreal representatives of the Neotropical spiny rats, family Echimyidae. They are found in the Atlantic forest of eastern Brazil, from the state of Ceará to Rio Grande do Sul, reaching the São Francisco and Paraná river basins in the west. Members of the genus *Phyllomys* are certainly among the most understudied representatives of the Neotropical mammal fauna. They have recently been referred to as *Nelomys* (e.g., Emmons and Feer, 1997; Thomas, 1916a), or frequently grouped with the Amazonian genus *Echimys*, as in the case of every recent systematic catalog (e.g., Cabrera, 1961; Ellerman, 1940; McKenna and Bell, 1997; Woods, 1993).

Nearly a century ago, Thomas (1916a, p. 297) pointed out that "With regard to the species of *Nelomys*, much confusion and ignorance exists, largely owing to the fact that so many of the earlier species were described without their exact localities being known, and often without reference to their dental characters." Unfortunately, this situation has not changed significantly since then. There are ten living described species (Emmons et al., 2002), and practically nothing is known of their natural history, ecology, and behavior. Their precise geographic limits are also unknown (Emmons and Feer, 1997), because most species are represented by just a few museum specimens, most of them collected more than 50 years ago (e.g., *P. thomasi*; see Olmos, 1997). The group has never been revised and the literature concerning its members is restricted to original descriptions, which are usually very terse. Consequently, museum specimens are often unidentified or misidentified. Most authors of recent catalogs and compilations simply had no opportunity to examine many, if any, representatives of this group (e.g., Cabrera, 1961; Tate, 1935).

Taxonomic History of *Phyllomys*

Tate (1935) reviewed the early taxonomic history of the species that can be assigned to *Phyllomys* under *Echimys*. There is still substantial chaos surrounding echimyid taxonomy (see Emmons et al., 2002), mainly as a historical consequence of the indiscriminate use of names, especially during the mid-nineteenth century, when international rules of nomenclature were not in use, and the French and German schools had different points of view (Tate, 1935).

The name *Nelomys* Jourdan, 1837, although older than *Phyllomys* Lund, 1839, is a junior synonym of *Echimys* Cuvier, 1809, and, therefore, it is not available (Emmons et al., 2002). Most recent authors (e.g., Cabrera, 1961; Ellerman, 1940; McKenna and Bell, 1997; Woods, 1993) erroneously attribute the name *Nelomys* to Cuvier. Cuvier's report, based on a memoir written by Jourdan, was presented to the French Academy of Sciences on 2 January 1838. Although the paper came out in the 1837 volume of the *Annales des Sciences Naturelles*, the date on the paper itself takes precedence and the correct citation should therefore be Cuvier (1838, not 1837). As members of the Academy, Cuvier and Duméril were in charge of introducing Jourdan and presenting his research, since he was not a member. Part of Jourdan's memoir, however, had already been published on 9 October 1837 in the *Comptes Rendus Hebdomadaires des Séances de L'Academie des Sciences* (vol. 15, date of publication printed on page 69), predating Cuvier's report. Judging by the remark "ms. ?" found on page 419 of Tate (1935), he was apparently unaware of the paper in the *Comptes Rendus*, and believed Jourdan's memoir was an unpublished manuscript.

Jourdan begins his description of *Nelomys* by stating: "Ce génre formé aux dépens du genre Échymys des auteurs, a pour type l'Échimys crsitatus [*sic*]" (Jourdan, 1837, p. 522). This statement makes *Echimys cristatus* Desmarest, 1817 the type of his new genus *Nelomys*. However, *Echimys cristatus* is a junior synonym of *Myoxus chrysurus* Zimmermann, 1780, which is the genotype of *Echimys* Cuvier, 1809, and therefore *Nelomys* becomes a junior synonym of *Echimys*. Jourdan's main goal in creating his new genus was to split the arboreal echimyids with short and round ears, short feet, and hairy tail, such as "*Echimys huppé*" (= *Echimys chrysurus*) and his "*Nélomys de Blainville*," from the terrestrial forms with long ears, long feet, and scaly tail, represented by "*Echimys de Cayenne*" (= *Proechimys cayennensis*) (Cuvier, 1838). His view is reflected in Geoffroy's (1840) monograph, where the terrestrial forms are listed under *Echimys* and the arboreal under *Nelomys*, except for *Dactylomys*, which is placed in its own genus.

The next available name for the Atlantic tree rats is *Phyllomys* Lund, first published in 1839 with a sentence describing the dentition in which the upper molars comprise four simple transverse laminae (Lund, 1839a, p. 226). This article in French is an extract based on three of Lund's monumental memoirs, which were published a year later in Danish (Lund, 1840a; for the explanation as to why the date 1840 is correct, see Musser et al., 1998, pp. 330–331). *Phyllomys* was treated on page 243 and illustrated in plate 21, figures 12 and 13. The figure captions reading "*Phyllostomus*" are obviously wrong as pointed out by Paula Couto (1950, p. 559).

Oldfield Thomas had a clear understanding of the distinctiveness of the genus *Phyllomys* and its contents, although he believed *Phyllomys* was a junior synonym of *Nelomys* (Thomas, 1916b). He showed that it is distinct from *Echimys* in that the upper molars of *Phyllomys* are modified into transverse plates, and from *Diplomys*, in which both upper and lower molars have transverse plates (Thomas, 1916b). Furthermore, he described two new species: *Phyllomys medius* from Paraná (Thomas, 1909) and *P. lamarum* from Bahia (Thomas, 1916a).

In his influential paper, Tate (1935) synonymized both *Nelomys* and *Phyllomys* under "*Echimys*," and most authors after him followed the same path (e.g., Cabrera, 1961; Ellerman, 1940; McKenna and Bell, 1997; Vieira, 1955; Woods, 1993), with few exceptions, such as Moojen (1952) and Emmons and Feer (1990, 1997).

For a number of years João Moojen was the curator of mammals at the Museu Nacional do Rio de Janeiro, Brazil, where he had access to the largest collection of *Phyllomys* specimens available, consisting of representatives of almost every species, including large series of some. He described a new species, *P. kerri* (Moojen, 1950), then considering *Phyllomys* as a subgenus of *Echimys*. In his classic book on Brazilian rodents (Moojen, 1952), however, he recognized *Phyllomys* as a full genus consisting of nine species. He called attention to the characters distinguishing *Phyllomys* from *Echimys*, such as upper molars composed of four transverse laminae, upper and lower molars relatively large, and frontal bones narrower in the former (Moojen, 1952, p. 137). He is the only modern author who used the name *Phyllomys*, and, judging by his mention of the true original description of *Phyllomys blainvilii* (Moojen, 1952, p. 139), he was probably already aware of the unavailability of *Nelomys* Jourdan, although he was not explicit about it.

When introducing the tree rats, Emmons and Feer (1990, p. 212) resurrected the name *Nelomys*: "All but one species in SE Brazil belong to a distinct group, the Atlantic tree rats, *Nelomys*, sometimes placed in the genus *Echimys* or *Phyllomys*." Their book, although a field guide and "not intended to solve systematic problems" (Emmons and Feer, 1990, p. 3), is an important reference because the taxonomy of the tree rats adopted by them reflects years of Emmons's own research, based on the examination of practically all museum specimens of this group, including the types (see Emmons et al., 2002). The text regarding the Atlantic tree rats is virtually the same in the second edition of their field guide (Emmons and Feer, 1997).

Emmons et al. (2002) reviewed the named forms of *Phyllomys* and clarified the taxonomy of the genus and its contents. They associated names with the correct entities, based on the examination of the type material available for each species and original descriptions. They designated lectotypes when no holotypes had originally been designated, restricted type localities, and recognized nine described species: *Phyllomys blainvilii, P. brasiliensis, P. nigrispinus, P. unicolor, P. dasythrix, P. thomasi, P. medius, P. lamarum,* and *P. kerri.* In addition, Emmons et al. (2002) described the new species *P. pattoni* based on specimens erroneously referred to as "*P. brasiliensis*" in the literature and museum records for over a century.

THE ATLANTIC FOREST

The Atlantic forest occurs along the east coast of South America, from northeasten Brazil to the southernmost border of this country, extending west into Argentina and Paraguay. It is one of the most endangered ecosystems in the world, with high levels of species diversity and endemism (Coimbra-Filho and Câmara, 1996; Costa et al., 2000; Dean, 1995; Fonseca, 1985; Morellato and Haddad, 2000; Mori et al., 1981; Mori, 1989; Por, 1992), and for that reason it has been considered one of the most important hotspots for conservation (Mittermeier et al., 1998; Myers et al., 2000). The Atlantic forest houses approximately 8,000 endemic plant and 567 vertebrate species (Myers et al., 2000). Approximately 230 species of mammals are found in the Atlantic forest, more than 70 of them endemic (Fonseca et al., 1999). Along with *Phyllomys*, other echimyids typically found in the Atlantic forest include the painted tree rat (*Callistomys pictus*), the bamboo rat (*Kannabateomys amblyonyx*), the terrestrial spiny rats (*Trinomys* spp.), and the semifossorial *Euryzygomatomys spinosus*.

It is estimated that only 6.8% of the forest cover that existed prior to European colonization remains today, scattered among highly fragmented forest patches (Conservation International do Brasil et al., 2000). In the present report, I refer to the Atlantic forest in a broad sense (*sensu lato*), including broadleaf evergreen rainforests found on the coast, montane mixed rainforests (*Araucaria* forests) in the south, and semideciduous forests inland, including the forest islands ("brejos") found in the dry areas in the northeast. On the coastal lowlands, the Atlantic forest extends as a narrow strip, bordered on the western margin by a mosaic of open vegetation types, including the Caatinga (thorn scrub forest) in the northeast, the Cerrado (Brazilian savannas) in central Brazil, and the Pampas (grasslands) in the south.

The climate varies from very humid to dry, mainly due to a rain shadow effect driven by the coastal mountains (Hueck, 1972; Mori, 1989). On the coast, the weather is very humid and precipitation is around 3000 mm per year, with some areas reaching over 4000 mm per year with virtually no dry season (Por, 1992). Inland areas west of the coastal mountains receive markedly less rain (1000–1500 mm/year) and the climate becomes more seasonal, with a marked dry season, usually lasting for five months. The topography is heterogeneous, ranging from flat lowland areas on the coast to steep hills and escarpments of the Serra do Mar and Serra da Mantiqueira reaching 1500–2000 m, with the highest peak, Pico da Bandeira, at 2890 m of elevation.

OBJECTIVES

The present study aims to increase understanding of the evolutionary history of the Atlantic tree rats of the genus *Phyllomys* by examining the variation of natural populations using diverse sources of genetic and morphological data. The specific goals are to clarify the taxonomy of the group, infer the phylogenetic relationships among species, understand the processes that lead to the present diversity and distribution, investigate the morphological variation within and among species, and discuss aspects of their natural history and conservation status.

CONCEPTUAL FRAMEWORK: SPECIES CONCEPTS

Since I am delimiting species boundaries and describing new species, the species concept used in the present report must be explicit. There has been continuous debate over the conceptualization of species, and at least 22 species concepts are in use today (Mayden, 1997). The Biological Species Concept (BSC) of Mayr (1942), where species are "groups of actually or potentially interbreeding natural populations which are reproductively isolated from other such groups", has dominated mammalogy in the twentieth century (Corbet, 1997). There are, however, several difficulties with the BSC (e.g., how to deal with allopatric, asexual, or parthenogenetic forms), and it presents fundamental obstacles to describing and interpreting patterns and processes of evolutionary differentiation (Cracraft, 1989).

  With the growing acceptance of cladistic methods, several species concepts have been developed with roots in Hennig's (1966) seminal work (for a recent review, see Wheeler and Meier, 2000). These phylogenetic species concepts shifted the focus from reproduction to genealogical relationships, since the ability to interbreed is viewed as a shared-primitive character and not of consequence in the recognition of species. The two predominant phylogenetic species concepts are quite distinct; one emphasizes diagnosability (e.g., Cracraft, 1983; Eldredge and Cracraft, 1980; Wheeler and Meier, 2000), while the other emphasizes monophyly or autapomorphies (e.g., Mishler and Donoghue, 1982; Mishler and Theriot, 2000; Queiroz and Donoghue, 1988; Rosen, 1978). In the version emphasizing diagnosis, species are basal taxa and must be diagnosed prior to phylogenetic analyses. Therefore, the terms monophyly, paraphyly, and polyphyly can only be applied above the species level because reticulation occurs below this level. In the version emphasizing monophyly, species must be monophyletic and phylogenetic relationships may occur below species, and the only way to detect where reticulation ends and divergence begins is by conducting a phylogenetic analysis.

  Here I follow the monophyly version of the phylogenetic species concept, and regard species as the least inclusive, geographically bounded, well-supported clades resulting from the phylogenetic analysis. The ranking decision takes into account diagnosability in combination with branch lengths (relative number of autapomorphies). I use only molecular data to infer phylogenetic relationships, but species must be diagnosable by morphological traits, so that they can be associated with museum specimens with no molecular data, including the name-bearing types. Branch lengths provide an estimate of the evolutionary uniqueness of lineages and are taken into account, especially given that some lineages are known from single specimens. These deserve specific recognition if the branch leading to the sister group is very long, indicating an old splitting event and the accumulation of evolutionary history, corroborated by obvious morphological diagnostic traits. I do not assess the taxonomic status of the species described in the literature for which there are no molecular data. They are diagnosable by morphological characters and I assume they are monophyletic, pending future cladistic analyses of morphological data and/or the availability of molecular data.

MATERIALS AND METHODS

SPECIMENS AND LOCALITIES

Specimens of *Phyllomys*

I examined 410 museum specimens of *Phyllomys* (listed under Species Accounts below) consisting of skins, skulls, post-cranial skeletons, and fluid-preserved material housed at the following institutions: Field Museum of Natural History [now Field Museum], Chicago, USA (FMNH); Museu de Biologia Mello Leitão, Santa Teresa, Brazil (MBML); Museu de Ciências Naturais, Fundação Zoobotânica do Rio Grande do Sul, Porto Alegre, Brazil (MCNFZB); Museu de História Natural Capão da Imbuia, Curitiba, Brazil (MHNCI); Museu Nacional, Rio de Janeiro, Brazil (MNRJ); Museum of Vertebrate Zoology, University of California, Berkeley, USA (MVZ); Museu de Zoologia da Universidade de São Paulo, São Paulo, Brazil (MZUSP); Museu de Zoologia da Pontifícia Universidade Católica do Paraná, Curitiba, Brazil (MZPUCPR); Coleção de Mamíferos do Departamento de Zoologia, Universidade Federal de Minas Gerais, Belo Horizonte, Brazil (UFMG); Coleção do Departamento de Sistemática e Ecologia, Universidade Federal da Paraíba, João Pessoa, Brazil (UFPB); Museu de Zoologia João Moojen de Oliveira, Universidade Federal de Viçosa, Viçosa, Brazil (UFV); National Museum of Natural History, Smithsonian Institution, Washington D.C., USA (USNM); Zoologische Museum und Institut für Spezielle Zoologie, Museum für Naturkunde der Humboldt-Universität zu Berlin, Berlin, Germany (ZMB). I also examined uncataloged specimens deposited in the above institutions but here identified by the collector's initials: G. A. B. da Fonseca (GABF), F. Kuminese (FK), A. Langguth (AL), P. P. de Oliveira (RBPDA), E. D. Rosal (EDR), M. Steindel (MS), and A. Ximenez (AX).

In addition, I had access to photographs, notes, drawings, and measurements taken by Louise Emmons, James Patton, Alfredo Langguth, Joel Clary, and Dieter Kock on 55 other specimens, including types, housed at the British Museum (Natural History) [now Natural History Museum], London, England (BMNH); Field Museum of Natural History [now Field Museum], Chicago, USA (FMNH); Muséum d'Histoire Naturelle de la Ville de Genève, Geneva, Switzerland (MHNG); Museé de Lyon, Lyon, France (ML); Muséum National d'Histoire Naturelle, Paris, France (MNHN); Senckenberg Museum, Frankurt, Germany;

(SMF); Naturhistorisches Museum, Vienna, Austria (NMW); and Universitäts Zoologisk Museum, Copenhagen, Denmark (UZMC).

Specimens used in the molecular analyses were collected in the field during several expeditions led by different research groups (see Acknowledgments) throughout eastern Brazil since 1991. Most of these specimens are still uncataloged, bearing the initials and field number of the following collectors: H. G. Bergallo (HGB), A. Christoff (AC), L. P. Costa (initials LC, LPC), L. Geise and R. Cerqueira (initials FS, IG), V. Fagundes (VF), Y. Leite (YL), A. Paglia (AP), and J. C. Voltolini (initials JCV, NSV). Specimens were collected using Sherman, Tomahawk, or locally made live traps placed either on the ground or in tree branches and vines. Some were collected during night censuses, using a .410-gauge shotgun loaded with dust shot (#12), and others were caught by hand from nests in tree hollows during the day. Voucher specimens were preserved as study skins and skulls or in fluid, following standard museum procedures. Liver samples were preserved in ethyl alcohol and/or liquid nitrogen for DNA analyses.

Other Echimyid Taxa

In the comparisons among genera, I examined the following specimens: *Echimys chrysurus*: Brazil—USNM 549594, 549595 (52 km SSW Altamira, right bank Rio Xingu, Pará). *Diplomys labilis*: Panama—USNM 460170, 503821 (El Aguacate), USNM 335740 (San Blas Armila, Quebrada Venado), USNM 116667 (San Miguel Island). *Makalata macrura*: Perú—MVZ 153636 (south side Río Cenepa, opposite Huampami, Depto. Amazonas), MVZ 153637 (Huampami, Río Cenepa, Depto. Amazonas), MVZ 157977 (La Poza, Río Santiago, Depto. Amazonas); Brazil—MVZ 190621 (Igarapé Nova Empresa, left bank Rio Juruá, Amazonas).

Gazetteer

I determined the geographic location (latitude and longitude) of 139 collecting localities of *Phyllomys*. Localities obtained with a global positioning system receiver are listed to the precision of seconds. Locality data listed as degrees and minutes only were obtained from gazetteers (NIMA, 1997; Paynter and Traylor, 1991; Vanzolini, 1992), topographic maps from IBGE (Instituto Brasileiro de Geografia e Estatística, Brazil, scale 1:25,000 or 1:50,000), and unpublished sources (e.g., field notes, museum catalogs). Distribution maps were produced using the software Arcview 3.0a for Macintosh (Environmental Systems Research Institute, Inc.). All localities are in Brazil, and they are listed below by state (abbreviations in parentheses) sequentially from north to south. Numbers correspond to the same numbered localities on the map, figure 1.

Ceará (CE): 1. Crato, 07°14'S 39°23'W; **2.** Sítio Anil, Crato, 07°14'S 39°23'W; **3.** Sítio Baixa do Maracujá, Crato, 07°14'S 39°23'W; **4.** Sítio Belo

Horizonte, Crato, 07°14'S 39°23'W; **5.** Sítio Caiano, Crato, 07°14'S 39°23'W; **6.** Sítio Grangeiro, Crato, 07°14'S 39°23'W; **7.** Sítio Macaúba, Crato, 07°14'S 39°23'W; **8.** Sítio Minador, Crato, 07°14'S 39°23'W; **9.** Sítio Quebra Primeira, Crato, 07°14'S 39°23'W; **10.** Sítio Santa Rosa (Santa Fé), Crato, 07°14'S 39°23'W; **11.** Sítio Serra Baixa do Maracujá, Crato, 07°14'S 39°23'W; **12.** Sítio Serra Bebida Nova, Crato, 07°14'S 39°23'W; **13.** Sítio Serra da Inga, Crato, 07°14'S 39°23'W; **14.** Sítio Serra dos Guaribas, Crato, 07°14'S 39°23'W; **15.** Sítio Serrinha, Crato, 07°14'S 39°23'W; **16.** Sítio Trindade, Crato, 07°14'S 39°23'W; **17.** Sítio Urucu de Fora, Crato, 07°14'S 39°23'W; **18.** Chapada do Araripe, 7 km SW Crato, 07°16'39"S 39°27'03"W 960 m.

    Paraíba (PB): **19.** Camaratuba, Mamanguape, 06°50'S 35°07'W; **20.** Uruba, Mamanguape, 06°50'S 35°07'W; **21.** João Pessoa, próximo ao NUPPA, 07°07'S 34°52'W.

    Pernambuco (PE): **22.** Reserva Ecológica Charles Darwin, Igarassu, 07°50'S 34°54'W; **23.** Dois Irmãos, Recife, 08°03'S 34°54'W; **24.** Sítio Cavaquinho, Garanhuns, 08°54'S 36°29'W.

    Alagoas (AL): **25.** Sítio Angelim, Viçosa, 09°24'S 36°14'W.

    Bahia (BA): **26.** Lamarão, about 70 miles [112 km] NW Salvador, 11°47'37"S 38°52'54"W 300 m; **27.** Várzea da Canabrava, Seabra, 12°10'S 41°39'W; **28.** 9 km SE Feira de Santana, 12°15'S 38°57'W; **29.** Fazenda Boa Vista, Feira de Santana, 12°15'S 38°57'W; **30.** Fazenda Estiva, Feira de Santana, 12°15'S 38°57'W; **31.** Fazenda Estrada Nova, Feira de Santana, 12°15'S 38°57'W; **32.** Fazenda Morro, Feira de Santana, 12°15'S 38°57'W; **33.** Fazenda Quituba, Feira de Santana, 12°15'S 38°57'W; **34.** Fazenda Salgado, Feira de Santana, 12°15'S 38°57'W; **35.** Fazenda Terra Nova, Feira de Santana, 12°15'S 38°57'W; **36.** Fazenda Três Riachos, Feira de Santana, 12°15'S 38°57'W; **37.** São Gonçalo, 30 km SW Feira de Santana, 12°25'S 38°58'W; **38.** Fazenda Santa Rita, 8 km E Andaraí, 12°48'06"S 41°15'41"W 399 m; **39.** 12 km de Lapa, Bom Jesus da Lapa, 13°15'S 43°25'W; **40.** Bom Jesus da Lapa, 13°15'S 43°25'W; **41.** Lapa, Bom Jesus da Lapa, 13°15'S 43°25'W; **42.** Rio São Francisco, Bom Jesus da Lapa, 13°15'S 43°25'W; **43.** Aritaguá-Urucutuca, Ilhéus, 14°39'28"S 39°07'31"W; **44.** Fazenda Almada, Ilhéus, 14°39'36"S 39°11'25"W; **45.** Itabuna, 14°48'S 39°16'W; **46.** São Felipe, 14°49'S 41°23'W; **47.** Fazenda Pirataquicê, Ilhéus, 14°50'S 39°05'W; **48.** Mata Fortuna, Itabuna, 14°57'S 39°19'W; **49.** Una, Ilhéus, 15°18'S 39°04'W; **50.** Mangue do Caritoti, Caravelas, 17°43'30"S 39°15'35"W 0 m; **51.** Fazenda Monte Castelo, Ilha da Cassumba, 7 km SW Caravelas, 17°48'06"S 39°15'49"W 32 m; **52.** Colônia Leopoldina (now Helvécia), 50 km SW Caravelas, 17°48'30"S 39°39'49"W 59 m; **53.** Helvécia, Nova Viçosa, 17°48'30"S 39°39'49"W.

    Espírito Santo (ES): **54.** Fazenda Santa Terezinha, 33 km NE Linhares, 19°08'S 39°57'W 50 m; **55.** Rio São José, 19°10'S 40°12'W; **56.** Povoação, Linhares, 19°37'S 39°49'W; **57.** Estação Biológica de Santa Lúcia, Santa Teresa, 19°57'S 40°31'W 600–900 m; **58.** Reserva Biológica de Duas Bocas, Cariacica,

20°17'S 40°28'W; **59.** Mata da Torre, Vitória, 20°19'S 40°21'W; **60.** Parque Estadual da Fonte Grande, Vitória, 20°19'S 40°21'W; **61.** Hotel Fazenda Monte Verde, 24 km SE Venda Nova do Imigrante, 20°20'S 41°08'W.

Rio de Janeiro (RJ): 62. Fazenda São José da Serra, 6 km E, 9.2 km N (by rd.) Bonsucesso, Sumidouro, 22°12'S 42°44'W 1000 m; **63.** Nova Friburgo, 22°16'S 42°32'W; **64.** Fazenda Alpina, Teresópolis, 22°25'S 42°50'W; **65.** Fazenda Comari, Teresópolis, 22°26'S 42°59'W 950 m; **66.** Teresópolis, 22°26'S 42°59'W; **67.** Fazenda Rosimery, Município de Cachoeiras de Macacu, 22°29'S 42°51'W; **68.** Fazenda União, Casimiro de Abreu, 22°29'S 42°12'W; **69.** Reserva Biológica de Poço das Antas, Silva Jardim, 22°31'S 42°17'W; **70.** FS 6, Município de Cachoeiras de Macacu, 22°32'S 42°48'W; **71.** FS 12, Município de Cachoeiras de Macacu, 22°31'S 42°47'W; **72.** Santa Cruz, estrada Rio-Petrópolis, 22°39'S 43°17'W; **73.** Monte São Francisco, Jacarepaguá, Rio de Janeiro, 22°55'S 43°21'W; **74.** Saco de São Francisco, Niterói, 22°55'S 43°06'W; **75.** Tijuca, Trapicheiro, Rio de Janeiro, 22°57'S 43°17'W; **76.** Ilha Grande, 23°09'S 44°14'W; **77.** Vila Dois Rios, Ilha Grande, 23°09'S 44°14'W.

Minas Gerais (MG): 78. Mocambinho, Jaíba, 15°06'S 44°03'W; **79.** Estação Ecológica de Acauã, 17 km N Turmalina, 17°08'S 42°46'W 800 m; **80.** Fazenda Santa Cruz, Felixlândia, 18°46'14"S 45°08'34"W; **81.** Serra do Ibituruna, Governador Valadares, 18°53'S 41°55'W; **82.** Lagoa Santa, 19°38'S 43°53'W; **83.** Sumidouro, 12 km NW Lagoa Santa, 19°32'28"S 43°56'28"W; **84.** Fazenda Montes Claros, Caratinga, 19°50'S 41°50'W; **85.** Mata Paraíso, Viçosa, 20°48'18"S 42°51'32"W 750 m; **86.** Silvicultura, Viçosa, 20°45'S 42°53'W; **87.** Alto da Consulta, Poços de Caldas, 21°48'S 46°34'W; **88.** Fazenda Paraíso, Além Paraíba, 21°52'S 42°41'W; **89.** Fazenda São Geraldo, Além Paraíba, 21°52'S 42°41'W; **90.** Fazenda do Bené, 4 km SE Passa Vinte, 22°14'S 44°12'W 680 m; **91.** Fazenda da Onça, 13 km SW Delfim Moreira, 22°36'S 45°20'W 1850 m.

São Paulo (SP): 92. Vanuire, 21°47'S 50°23'W; **93.** Teodoro Sampaio, 22°31'S 52°10'W; **94.** Piquete, 22°36'S 45°11'W; **95.** Itatiba, 23°00'S 46°51'W; **96.** Parque Estadual da Serra do Mar, Núcleo Santa Virgínia, 10 km NW Ubatuba, 23°21'30"S 45°07'30"W 850 m; **97.** Estação Experimental, Ubatuba, 23°25'00"S 45°06'47"W; **98.** Ubatuba, 23°26'S 45°04'W; **99.** Floresta Nacional de Ipanema, 20 km NW Sorocaba, 23°26'07"S 47°37'41"W 550–970 m; **100.** São Paulo, 23°32'S 46°37'W; **101.** near São Paulo, 23°32'S 46°37'W; **102.** Itapetininga, 23°35'S 48°03'W; **103.** Taboão da Serra, 23°36'S 46°46'W; **104.** Estação Biológica de Boracéia, Salesópolis, 23°39'S 45°54'W; **105.** Interlagos, São Paulo, 23°43'S 46°42'W; **106.** Rio Guaratuba, Santos, 23°45'S 45°55'W; **107.** Ilhabela, Ilha de São Sebastião, 23°47'S 45°21'W; **108.** Ilha de São Sebastião, 23°50'S 45°18'W; **109.** Trilha da Água Branca, Ilha de São Sebastião, 23°50'S 45°18'W; **110.** Primeiro Morro, 24°18'S 47°44'W; **111.** Barra do Ribeirão Onça Parda, 24°19'S 47°51'W; **112.** Ribeirão Fundo, 24°20'S 47°45'W; **113.** Barra do Rio Juquiá, 24°22'S 47°49'W; **114.**

Barra de Icapara, 24°41'S 47°28'W; **115.** Ilha do Cardoso, Cananéia, 25°08'S 47°58'W.

Paraná (PR): 116. Porto Camargo, Rio Paraná, 23°21'S 53°43'W; **117.** Rio Paracaí, 23°41'S 53°57'W; **118.** Salto Morato, Guaraqueçaba, 25°17'S 48°21'W; **119.** Parque Barigüi, Bairro Mercês, Curitiba, 25°24'56"S 49°18'03"W 861 m; **120.** Roça Nova, Serra do Mar, 25°28'19"S 49°00'50"W 1000 m; **121.** Guajuvira, 25°36'S 49°32'W; **122.** Palmira, 25°42'S 50°09'W; **123.** Represa de Foz do Areia, 35 km S Pinhão, 26°00'S 51°30'W.

Santa Catarina (SC): 124. Joinville, 26°18'S 48°50'W; **125.** Lagoa da Conceição, Freguesia, Florianópolis, 27°33'S 48°27'W; **126.** Florianópolis, 27°35'S 48°34'W; **127.** Ilha de Santa Catarina, 27°36'S 48°30'W; **128.** Sítio Garça Branca, Rod. SC401, km 10, Florianópolis, 27°36'S 48°30'W; **129.** Praia dos Açores próximo a Pântano do Sul, Ilha de Santa Catarina, 27°46'S 48°30'W; **130.** Serra do Tabuleiro, 27°50'S 48°47'W.

Rio Grande do Sul (RS): 131. Usina Hidrelétrica de Itá, 27°16'S 52°19'W; **132.** Parque Nacional dos Aparados da Serra, Cambará do Sul, 29°15'S 49°50'W 800 m; **133.** São Francisco de Paula, 29°27'S 50°35'W; **134.** Pinheiros, Candelária, 29°47'S 52°44'W; **135.** Banhado do Pontal, Triunfo, 29°56'S 51°43'W; **136.** Porto Alegre, 30°04'45"S 51°07'30"W; **137.** Bairro Agronomia, Porto Alegre, 30°04'45"S 51°07'29"W 46 m; **138.** Itapuã, Viamão, 30°16'S 51°01'W; **139.** Parque Estadual de Itapuã, Viamão, 30°24'S 50°58'W.

FIGURE 1. Map of eastern Brazil showing collecting localities of *Phyllomys* spp. Numbers correspond to numbered localities in the Gazetteer.

MORPHOLOGY

Age Categories

I use the following abbreviations for the upper cheekteeth, unless otherwise specified: PM = pre-molar, M1 = first molar, M2 = second molar, M3 = third molar. I define the following ten age categories for the specimens, based on toothwear pattern (Fig. 2): (1) PM erupted and worn, M1 erupted but unworn; (2) M1 worn, M2 undeveloped; (3) M2 developed but unerupted; (4) M2 erupted but unworn; (5) M2 worn, M3 undeveloped; (6) M3 developed but unerupted; (7) M3 erupted but unworn; (8) M3 worn, PM–M2 with 9–12 independent laminae; (9) PM–M2 with 5–8 independent laminae; (10) PM–M2 with 0–4 independent laminae. For a tooth to be considered erupted, the occlusal surface had to be clearly above the bony socket. A tooth was considered unworn when no dentine was visible, and undeveloped when there was no visible sign of tooth. Laminae were considered independent when dentine of one loph was not connected with dentine from another loph.

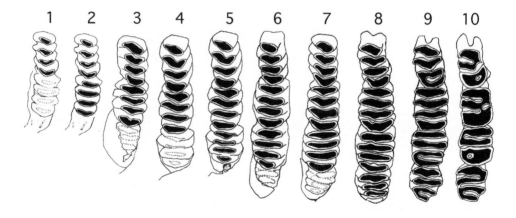

FIGURE 2. Right maxillary toothrows of *Phyllomys* showing age categories defined by tooth eruption and occlusal wear (see text for details).

Morphometric Analyses

External body measurements (in millimeters) and weight (in grams) listed are from museum tags or field notes, except when noted. I also took external measurements from study skins and fluid preserved specimens when no field measurements were available or when it was unclear how they were taken (e.g., hindfoot with or without claw).

I took the below 23 cranial measurements with a digital caliper to the nearest 0.01 mm (Fig. 3). These were based on Patton and Rogers (1983) and Silva (1998).

GSL	Greatest skull length, anterior-most projection of nasals to posterior-most projection of occipital region on midline of skull
NL	Nasal length, greatest length of nasals at midline
RL	Rostral length, diagonal measure from anterior edge of orbit lateral to lacrimal to anterior edge of nasals at midline
OL	Orbital length, greatest length of orbits
RB	Rostral breadth, breadth of rostrum at the suture between premaxilla and maxilla
IOC	Interorbital constriction, least distance between bony orbits
MB	Mastoid breadth, distance across cranium at mastoid processes
ZB	Zygomatic breadth, maximum width across outside margins of zygomatic arches
CIL	Condyloincisive length, anterior edge of upper incisors to posterior-most projection of occipital condyle
BaL	Basilar length, posterior margins of upper incisors to anterior edge of foramen magnum
D	Diastema length, posterior alveolar margin of upper incisors to anterior alveolar edge of PM4
MTRL	Maxillary toothrow length, anterior alveolar edge of PM4 to posterior alveolar edge of M3
Pla	Palatal length a, midline distance between posterior margins of upper incisors to anterior margin of mesopterygoid fossa
PLb	Palatal length b, anterior edge of PM4 to anterior edge of mesopterygoid fossa
IFL	Incisive foramina length, length of opening of foramina
BuL	Bullar length, maximal distance from anterior to posterior edges of tympanic bulla
PPL	Postpalatal length, posterior margin of inner aspect of zygomatic arch to a line perpendicular and tangent to greatest projection of occipital region
MPF	Mesopterygoid fossa width, maximum width taken at the suture between palatine and pterygoid bones

MaxB   Maxillary breadth, greatest breadth of maxilla on outside of M3

OccW   Occipital condyle width, width across outside margins of the occipital condyles

RD   Rostral depth, depth of rostrum at the suture between premaxilla and maxilla

CD   Cranial depth, vertical distance from ventral margin of bulla to top of cranium

CDM1   Cranial depth at M1

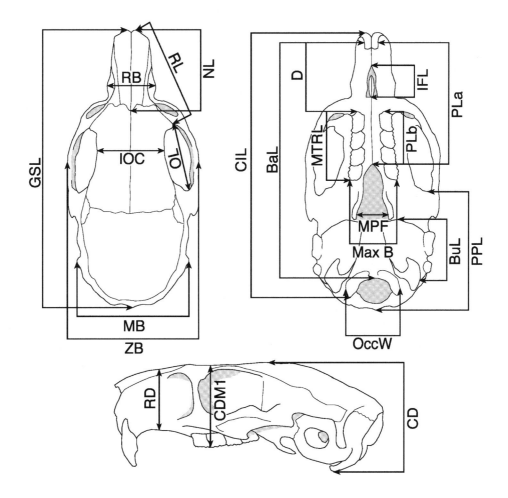

FIGURE 3. Position of 23 cranial measurements taken on skulls of *Phyllomys* (see text for abbreviations and explanation).

Cranial measurements were analyzed using univariate and multivariate statistical methods. Basic descriptive statistics (mean, standard deviation, range, and sample size) are given for all measurements. Two-way analyses of variance (ANOVA) were used to examine non-geographic variation (due to sex and age, as estimated from toothwear categories). Principal components analyses (PCA) were used to examine how individual measurements contribute to the total pool of variation, and discriminant function analyses (DFA) were used to allocate samples to a given species. All multivariate analyses used $\log_{10}$ transformed variables. All analyses were done using the software StatView 5.0 (SAS Institute, Inc.), except the DFA, performed in Statistica 4.1 (StatSoft, Inc.) for Macintosh.

KARYOTYPE

Cell suspensions were prepared in the field following a slightly modified version of the colchicine-hypotonic citrate protocol described by Patton (1967). Animals were injected intraperitonially with colchicine (0.01 ml/g body weight, using a 0.05% solution) one to two hours prior to sacrifice. During specimen preparation, the bone marrow was flushed from the shaft of the tibia with a hypotonic solution (KCl) and filtered through cheesecloth. After 15 minutes cells were centrifuged for 1–2 minutes using a hand-driven centrifuge and fixed with a 1:3 acetic acid-methanol solution (Carnoy's fixative). Cells were resuspended and fixed 3 more times and stored in cryogenic vials at room temperature. In the laboratory, samples were subsequently stored in a -20°C freezer, and slides were prepared 2–4 years later. First I resuspended the cell in fresh fixative and then prepared slides using the air-dry technique, stained them with Wright's stain, and examined and photographed them under oil immersion using bright-field optics.

MOLECULAR DATA AND PHYLOGENETIC ANALYSES

I extracted DNA using either the Chelex (BioRad) method or the Dneasy extraction kit (Quiagen, Inc.). Two fragments of mitochondrial cytochrome b (cyt b) gene (1140 bp) were amplified in 25 μl polymerase chain reactions (PCR) using the following primer pairs (5' to 3'):

```
MVZ   05    (CGAAGCTTGATATGAAAAACCATCGTTG)
MVZ   16    (AAATAGGAARTATCAYTCTGGTTTRAT)
MVZ   127   (TRYTACCATGAGGACAAATATC)
MVZ   108   (CCAATGTAATTTTTATAC)
```

The temperature profile used in the PCR reactions was initial denaturation 94°C/5 min., then 39 cycles of 94°C/30 sec., 48°C/30 sec., 72°C/ 45 sec., and final extension at 72°/45 sec.

After an agarose gel check, samples were cycle-sequenced using the ABI-Prism d-Rhodamine kit (Applied Biosystems, Inc.) through 25 cycles of 95°C/30 sec., 50°C/15 sec., 60°C/4 min using the same primers listed above. Sequences were obtained using an ABI Prism 377 (Applied Biosystems, Inc.) automated sequencer, and were aligned by eye using the software Sequencher 3.0 (Gene Codes Corp.) in a Macintosh computer.

Phylogenetic analyses were performed with PAUP*4.0 (Swofford, 2000), using maximum parsimony and maximum likelihood as optimality criteria. Heuristic searches were conducted utilizing the tree-bisection-reconnection (TBR) algorithm, via random stepwise addition with ten replicates (except for maximum likelihood), and collapsing zero length branches. For parsimony, character states were optimized using the accelerated transformation (ACCTRAN) option. For maximum likelihood, I used the program Modeltest 3.0.4 (Posada and Crandall, 1998) to select the most appropriate model of molecular evolution through a nested likelihood ratio test. Maximum likelihood analyses were then performed in two steps. First, the strict consensus of the most parsimonious trees was scored for maximum likelihood parameters, using the appropriate model. The resulting scores were then input in a heuristic search for the maximum likelihood tree.

Measures of support employed in tree comparisons were the ensemble consistency and retention indices (CI and RI, respectively). The CI was computed excluding uninformative characters. Support for clades within trees was assessed using the Bremer support, or decay, index (Bremer, 1988), calculated using the program Autodecay 4.0 (Eriksson, 1998), and bootstrap analysis (Felsenstein, 1985). For parsimony, 100 bootstrap replicates were performed using full heuristic searches, each with ten replicates and random addition sequence of taxa. For maximum likelihood, 100 bootstrap replicates were conducted via "fast" stepwise-addition.

GENETIC STRUCTURE

The genetic structure was investigated using the algorithms implemented in the software Arlequin 2.000 (Schneider et al., 2000), except for the tabulated pairwise genetic distances, which were calculated in PAUP, using the Kimura 2-parameter model (Kimura, 1980). To test for selective neutrality, I used Tajima's D (Tajima, 1989), which is based on the infinite sites model, and the statistical significance was calculated assuming a beta-distribution limited by maximum and minimum possible D values, as implemented in Arlequin. Minimum spanning networks (see Excoffier and Smouse, 1994) among haplotypes were constructed using a distance matrix based on pairwise differences. The distribution of the observed number of differences between pairs of haplotypes (mismatch distribution, Rogers and Harpending, 1992; Rogers, 1995) was also estimated using pairwise differences and compared with an estimated stepwise expansion model. For this comparison I used the sum of squared deviations (SSD) between the observed and expected mismatch and Harpending's Raggedness index (HRi) as implemented in Arlequin.

NATURAL HISTORY AND HABITAT DATA

Notes on the method of capture and reproductive condition of museum specimens were also taken from skin tags and associated records. The collecting date was used in conjunction with the toothwear age category and presence of embryos to estimate reproductive season. I investigated diet by dissecting individuals and analyzing the morphology of the digestive tract. Habitat information is briefly summarized from secondary sources for some of the collecting localities. Given the confusion surrounding the scientific terms related to canopy biology, I follow Moffett (2000).

RESULTS AND DISCUSSION

TWO NEW SPECIES OF *PHYLLOMYS*

During a mammal survey of the Atlantic forest (Projeto Inventário Faunístico da Mata Atlântica, Fundação Biodiversitas; see Acknowledgments) carried out between 1991 and 1992, we collected three specimens of *Phyllomys* in southeastern Brazil, belonging to three species. One of these corresponds to specimens traditionally referred to as "*Phyllomys brasiliensis*" in the literature and museum collections. After examining museum specimens, including type material of every species, we concluded that nearly all specimens formerly identified as *Phyllomys brasiliensis* in fact represented a distinct and, at that time, an undescribed species, which we subsequently named *Phyllomys pattoni* (Emmons et al., 2002). The two other species are described immediately below. Each represents a unique lineage, identified in the phylogenetic analysis (see Molecular Phylogenetics, below), but I name and describe them first, as follows.

Phyllomys lundi, sp. nov.

Holotype: MNRJ 62392, adult male collected by Yuri Leite on 12 October 1991 (field number YL 7). It consists of a study skin, skull, carcass fixed in formalin and preserved in 70% ethanol, liver sample in 95% ethanol.

Type locality: Fazenda do Bené, 4 km SE Passa Vinte, Minas Gerais, Brazil, 22°14'S 44°12'W 680 m (Fig. 1, locality 90).

Diagnosis: One of the smallest species in the genus. External body measurements of the holotype: total length = 413 mm; tail length = 204 mm; hindfoot length = 36 mm; ear length = 16 mm; weight = 145 grams. Pelage dominantly orange intermixed with black. Neck and thighs markedly orange. Spines conspicuous from the neck to the tail. Ventral hairs cream with a white base, giving a washed aspect. Tail brown, hairy from base to tip. Forefeet covered with brown-yellow hairs, except fingers, gray-white. Hindfeet covered with golden-creamy hairs; toes with silver hair. Skull delicate, rostrum relatively narrow and long (Fig. 4). Interorbital region wide and convex. Only three transverse plates in upper M3. Mandible with a short coronoid process and a shallow sigmoid notch.

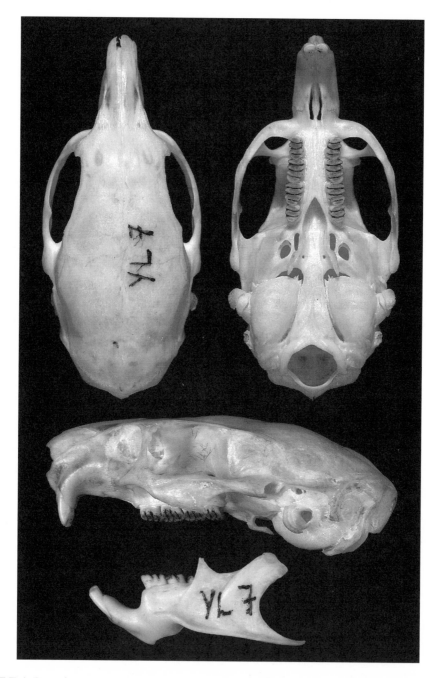

FIGURE 4. Dorsal, ventral, and side views of the skull and side view of the mandible of the holotype of *Phyllomys lundi*, MNRJ 62392 (field number YL 7). Magnification = x2.

Comparisons: *Phylomys lundi* differs from most species in the genus by the combination of its small size and orange-tipped aristiforms. Other small to medium species with orange-tipped aristiforms are *P. blainvilli* and *P. lamarum*. Compared to *P. blainvilii, P. lundi* tends to be smaller, has a darker fur, shorter aristiforms with a shorter orange band near the tip (Fig. 5), weaker supraorbital ridges, and narrower palate. Its darker coat color, hairy tail, and weakly developed supraorbital ridges distinguish *P. lundi* from *P. lamarum*. Given that *P. lundi* is one of the smallest species in the genus, its phallus is relatively long, differing from those of all other taxa by having a cylindrical rather than tapering distal end (Fig. 6).

Referred specimens: The only other museum specimen that can be assigned to this species was collected at Reserva Biológica de Poço das Antas, Silva Jardim, Rio de Janeiro (Fig. 1, locality 69), by Paula Procópio de Oliveira and Yuri Leite (eartag number 2228) on 4 April 1990. It is an adult female preserved as a study skin only, measuring: head and body length = 184 mm; tail length = 200 mm; hindfoot length = 38 mm; ear length = 17 mm; weight = 202 grams. The picture in Nowak (1991, p. 951) labeled *"Echimys nigrispinus"* depicts an individual of the same species, captured by Jody R. Stallings in a patch of exotic bamboo (*Bambusa tuldoides*) at Poço das Antas in 1988.

Etymology: Named after Peter Wihelm Lund (1801–1880), the Danish naturalist who lived and worked in Lagoa Santa, Minas Gerais, Brazil, during the nineteenth century. He assembled one of the most important mammal collections from a single locality in the Neotropics, and made outstanding contributions towards describing the Pleistocene and Recent mammal faunas of Brazil, including the description of the genus *Phyllomys*.

Description: This is a small-sized and heavily spined *Phyllomys*. The dorsal pelage is blackish brown intermixed with orange. The spines are conspicuous from the base of the tail to the neck. The aristiforms on rump average 22 mm in length and 0.8 mm in maximum width, with dark edges, gray at the base, gradually darkening towards the orange tip (Fig. 5E). Setiforms on rump are shorter than aristiforms. The dorsal portion of neck and outer thighs are markedly orange. The ventral region is covered with cream-brown white-based hair. Pure white patches are present on the chin, throat, axillary, and inguinal regions. The transition between the darker dorsum and the lighter-colored venter is sharp. The tail is approximately as long as head and body, dark brown above and light brown below, hairy but scales are visible, and hairs are longer near the tip. The nose is light orange, and mystacial vibrissae are black and long (65 mm). The longest supracilliary and genal vibrissa reach 45 mm. The pinna is nearly naked, sparsely covered with 15-mm-long, very thin hair, especially along the edges. Short (10 mm) aristiforms are abundant near the tragus. The dorsal surface of the forefeet is dark brown, contrasting with silver-orange fingers; the digital tufts extend beyond the foreclaws for less than 1 mm. The ventral surface has three equidistant and equal-sized interdigital pads. The hindfeet are silver-orange above, and the digital tufts extend beyond the hindclaws for 1–2 mm. Four equidistant interdigital pads are present.

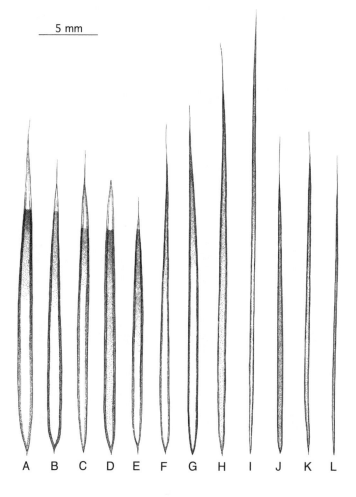

FIGURE 5. Aristiform hairs from the rump of 12 species of *Phyllomys*: A—*P. brasiliensis* (AP 48); B—*P. lamarum* (LC 73); C—*P. balainvilii* (LPC 290); D—*P. pattoni* (MNRJ 62391, the holotype); E—*P. lundi* (MNRJ 62392, the holotype); F—*P. nigrispinus* (MZUSP 1031); G—*P. kerri* (MNRJ 5463); H—*P. thomasi* (MZUSP 3198); I—*P. medius* (MNRJ 48864); J—*P.* aff. *dasythrix* (AC 640); K—*P. dasythrix* (MCNFZB 46); L—*P. mantiqueirensis* (MNRJ 62393, the holotype). Distal part pictured in white is actually orange/ochraceous in A, B, C, D, and E.

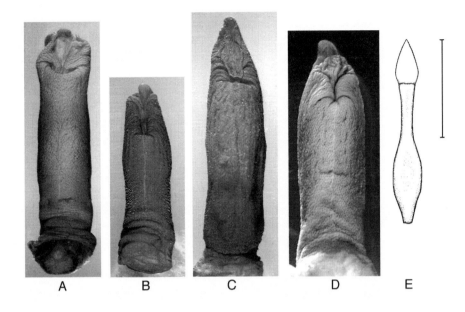

FIGURE 6. Ventral view of the glans penis of: A—*P. lundi* (MNRJ 62392, the holotype); B—*P. mantiqueirensis* (MNRJ 62393, the holotype); C—*P. medius* (EDR 8); D—*P. pattoni* (YL 198); and E—baculum of *P. pattoni* (YL 199), including cartilagenous tip, is shown to the right. Scale bar = 5 mm.

The glans penis is cylindrical, 10 mm in length and 2.8 mm in maximum width, with a slight subterminal narrowing (Fig. 6A). The opening of the intromittent sac is circular, bordered by two large lateral skin folds that extend into the sac. The intromittent sac is deep (ca. 4 mm), with no spikes on its floor. The urethral lappet is wrinkled but is coated by a smooth epidermis.

The skull appears small and delicate; the interorbital region is wide, slightly concave, diverging posteriorly with a small post-orbital process. The supraorbital ridges are well developed, extending posteriorly to the level of the posterior margin of the squamosal root of the zygomatic arch. The rostrum is relatively narrow and elongated. The incisive foramina are long (length approximately 1/2 of diastema), bullet-shaped, with two lateral and one medial posterior ridges extending to the anterior palate. The palate is not especially wide, and palatine width is approximately the same as tooth width at M1. The upper toothrows are slightly divergent posteriorly. Upper M3 has only three transverse plates and upper incisors are orthodont. The mesopterygoid fossa is wide, forming an angle of approximately 55 degrees and reaching the third lamina in M2. The squamosal does not contribute to the post-orbital process of the zygoma. Mastoid process reaches the midline of the external auditory meatus. Ventral root of the angular process is deflected laterally. The coronoid process is short, not reaching the height of the condyle, and projects upward and slightly forward. The sigmoid notch is very shallow. Selected skull measurements are given and discussed under Cranial Morphometrics, below.

Habitat: The habitat at the type locality is old, second growth rainforest with a dense overstory, where most trees are 20 cm in diameter (DBH), and the largest ones reach almost 1 m in DBH with crowns 20 m in height. The forest patch runs along the Rio Preto at 680 m elevation. It is on private property. The terrain is moderately steep (30°–60°), with rock outcrops and relatively few fallen logs on the ground, a sparse understory usually reaching 1.5 m of height, few lianas and vines, some arboreal bromeliads, abundant arboreal lichens, and many palm and fruit trees. The type specimen was collected in a Tomahawk trap placed 1 m high on a tree branch, less than 10 m from a small creek. Other small mammals collected in this area during the same period were the echimyid *Trinomys gratiosus* and the marsupials *Didelphis aurita* and *Philander frenatus*.

The other known locality for this species, Poço das Antas (ca. 5,000 ha; Fig. 1, locality 69) is 200 km ESE of the type locality. The annual rainfall is about 2000 mm and the average temperature was 24°C between November 1989 and October 1990 (Leite et al., 1996). The specimen (RBPDA 2228) was caught in a Sherman trap placed 1.5 m high on a vine climbing a tree trunk in the forest site locally known as Portuense. The habitat at the trap station falls into the hilltop forest category of Dietz et al. (1997), who studied golden lion tamarins (*Leontopithecus rosalia*) at the Portuense forest. According to those authors, this habitat is characterized by tall, selectively logged mature forest on dry ridge-top soils, overstory discontinuous with emergents reaching 32 m, sparse understory, and lack of discernible boundaries between the forest strata. Other characteristics of the Portuense forest include an average tree height of 10.1 m and diameter (DBH) of 15.7 cm, tree density of 99.6 individuals/0.1 ha, and relatively low density of lianas (1.49 on a scale of 1 to 10), medium density of tree hollows (39.8/ha), low bromeliad density (11.8/0.1 ha), dense overstory cover (7.94 on a scale from 1 to 10), and high tree species richness (0.431 trees species/number of trees) (Dietz et al., 1997).

Remarks: *Phyllomys lundi* is a unique entity, distinguished by molecular autapomorphies and diagnostic morphological characters, and its phylogenetic position within the genus is not clear. The phylogenetic analysis suggests that it is either related to *P. nigrispinus* or to the northeastern clade, with low support in either case (see Molecular Phylogenetics, below).

Phyllomys mantiqueirensis, sp. nov.

Holotype: MNRJ 62393, adult male collected by Yuri Leite on 17 November 1991 (field number YL 23). It consists of a study skin, skull, carcass fixed in formalin and preserved in 70% ethanol, liver sample in 95% ethanol.

Type locality: Fazenda da Onça, 13 km SW Delfim Moreira, Minas Gerais, Brazil, 22°36'S 45°20'W 1850 m (Fig. 1, locality 91).

Diagnosis: A small, soft-furred *Phyllomys* with a hairy tail tufted at the tip. External body measurements of the holotype: total length = 433 mm; tail length = 216 mm; hindfoot length = 41 mm; ear length = 18 mm; weight = 207 grams. Body brown-gray and fur soft and dense throughout. Tail as long as head and body, darker than body, covered with brown hair that gets denser and longer towards the tip, forming a long (30 mm) terminal tuft. Interorbital region narrow, lacrimal process well developed. Rostrum short and broad, incisive foramina small and oval-shaped. Mandible with a slim and long coronoid process and a deep sigmoid notch (Fig. 7).

Comparisons: *P. mantiquerensis* differs from most species of *Phyllomys* by having a soft fur, and therefore can only be confused with *P. dasythrix*. Its tufted tail, long mastoid process, lack of postorbital process, and laterally deflected ventral root of the angular process distinguish *P. mantiqueirensis* from *P. dasythrix*. The phallus of *P. mantiqueirensis* differs from that of other species of *Phyllomys* by the absence of urethral lappet (see Fig. 6), but the phallus of *P. dasythrix* has not been described yet.

Etymology: Named after the mountain range where the holotype was collected, Serra da Mantiqueira, which runs parallel to the coastal range (Serra do Mar) on the border of the states of Minas Gerais, Rio de Janeiro, and São Paulo in southeastern Brazil.

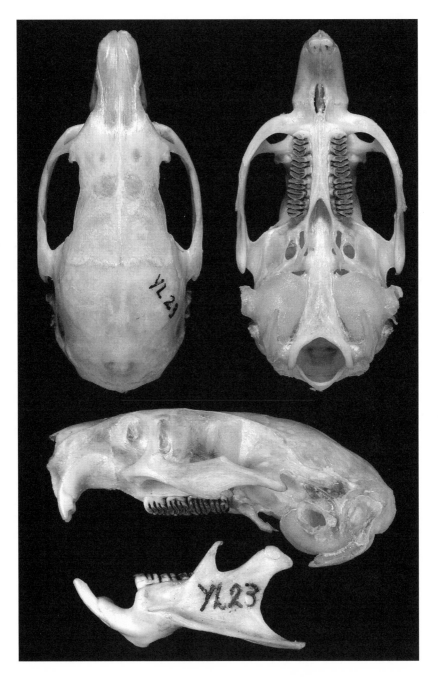

FIGURE 7. Dorsal, ventral, and lateral views of the skull and lateral view of the mandible of the holotype of *Phyllomys mantiqueirensis*, MNRJ 62393 (field number YL 23). Magnification = x2.

Description: This is a small, soft-furred *Phyllomys*. The dorsal pelage is light brown streaked with black. Sides of the body are lighter than the dorsum, grading to brown-yellow ventrally. Dorsal hair is white at the base, darkening towards the middle. Aristiforms on rump are black distally, averaging 24 mm long and 0.1 mm wide. Setiforms are abundant and similar to aristiforms, but shorter, with tip brown-yellow. Ventral hair is less than 10 mm long, white at the base and light cream-yellow at the tip. White patches are present on the throat, axillary, and inguinal regions. The head is slightly darker than the rest of the body, but the light cream muzzle appears paler than the rest of the face. Mystacial vibrissae emerge in pairs from a single pore, and including a longer and thicker brown vibrissa, and a shorter and thinner black one, longest reaching 70 mm. The longest supracilliary vibrissae reach 35 mm, and longest genal reach 40 mm. The ears are covered with thin and long (10 mm) blackish brown hair. The tail is long and dark brown contrasting with the lighter color of the body. Tail is covered with extension of body hair proximally for 30 mm, then it is well haired to the tip, with hidden scales. Tail hairs are short (5–7 mm) near the base, gradually becoming longer, reaching 30 mm at the tip, where they form a conspicuous tuft.

The glans penis is cylindrical proximally, with a slight medial swelling (Fig. 6B), tapering towards the tip, where the baculum forms a 1-mm-long pointed papilla. Length of glans is 8 mm, and maximum width is 2.5 mm. There is no urethral lappet, and the urethra opens directly into the intromittent sac, ca. 2 mm below the tip of the bacular papilla. The intromittent sac is shallow (ca. 2 mm), with no spines on its floor. The ventral margin of sac is U-shaped, lateral borders with several small folds.

The skull is delicate without pronounced crests and ridges, apart from the lambdoidal and occipital. Supraorbital ridges are present but weakly developed, extending to the fronto-parietal suture. Interorbital region is narrow and slightly divergent posteriorly without a pronounced postorbital process. The lacrimal process is very well developed. The rostrum is short and broad. The incisive foramina are short (length approximately 1/3 of diastema) and ovate, with two well-developed lateral and one medial posterior ridges, extending to the anterior palate. Upper toothrows are slightly divergent posteriorly. Cheekteeth are wide, palatine width less than width of M1. Upper incisors are orthodont with pale orange enamel. The mesopterygoid fossa is wide, forming an angle of approximately 45 degrees and reaching the posterior lamina in M2. The periotic capsule of mastoid is inflated and well developed. The post-orbital process of the zygoma is spinose, formed mainly by squamosal. The mastoid process extends to the inferior border of the external auditory meatus. The mandible is robust, and the ventral root of the angular process is laterally deflected. The coronoid process is slim but well developed, standing higher than condyloid process, forming a very deep sigmoid notch. Selected skull measurements are given and discussed in section Cranial Morphometrics, below.

Habitat: The type locality is a cool, wet montane forest on a hillside at 1850 m elevation. This is the highest elevation recorded for any species in the genus *Phyllomys*. The overstory is open and low (about 10 m), and trees usually have a diameter (DBH) of 15 cm; the largest are emergent araucarias (*Araucaria angustifolia*) that attain 70 cm DBH. The topography is moderately steep (30°–60°), with some fallen logs and no rock outcrops. There are few arboreal or terrestrial bromeliads, the understory is moderately dense, characterized by 1.5 m high shrubs, lianas, vines, abundant bamboos and lichens, and some ferns. The area is part of a Brazilian army camp. The holotype was collected in a Tomahawk trap placed 2 m high on a liana connecting two trees, 45 m from a small creek. Other small mammals collected in this area during the same period were the sigmodontine rodents *Akodon serrensis*, *Thaptomys nigrita*, and *Delomys dorsalis*.

Remarks: Known only from the holotype. This is a distinctive species, very different from all others in terms of both morphology and cyt b. Its phylogenetic position is not clear, but it seems to represent one of the oldest lineages within the genus, perhaps the sister group to all other extant forms (see Molecular Phylogenetics, below).

MOLECULAR PHYLOGENETICS

The entire cytochrome b gene (1140 bp) was sequenced for 34 individuals collected at 24 localities throughout the range of *Phyllomys* (Fig. 8), including eight of thirteen species recognized here (see Table 1). Missing from this data set are *P. unicolor*, known only from the holotype collected in the nineteenth century; *P. kerri*, known from 3 specimens of the type series; *P. thomasi*, endemic to the island of São Sebastião; and *P. medius*. Two sequences, LPC 246 and CIT 1357, were further excluded from the analysis since they were identical to those of LPC 290 and CIT 1346, respectively. Sequence for an additional specimen (MNRJ 43810 = LMP 27) was available from Lara et al. (1996). I used *Echimys chrysurus*, *Makalata macrura*, and *M. didelphoides* as outgroups based on previous analyses showing they are the closest relatives to *Phyllomys* (Lara et al., 1996). The final matrix had 38 OTUs and 1140 characters, 728 constant and 324 parsimony informative.

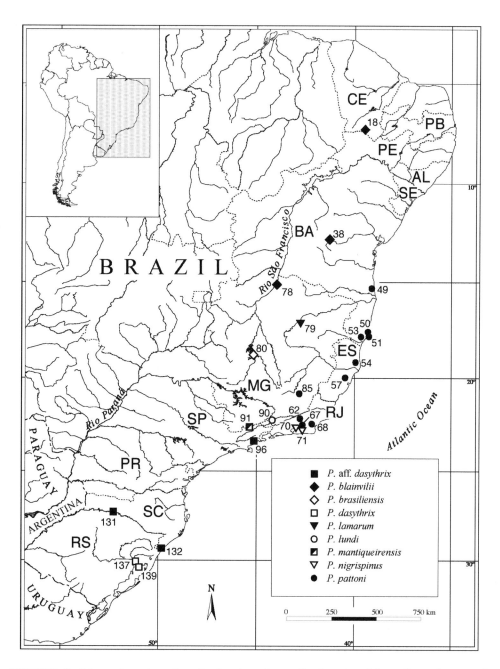

FIGURE 8. Map of eastern Brazil showing the collecting localities of specimens of *Phyllomys* used in the molecular analysis. See Table 1 for specimens.

TABLE 1. Specimens of *Phyllomys* used in the molecular phylogenetic analysis. The locality numbers correspond to the Gazetteer and map in figure 8.

Species	State: locality (locality number)	Specimen numbers
P. aff. dasythrix	RS: Parque Nacional dos Aparados da Serra (132)	AC 632
	RS: Usina Hidrelétrica de Itá (131)	CIT 1344, 1346, 1357
	SP: Parque Estadual da Serra do Mar (96)	NSV 160599
P. blainvilii	BA: Fazenda Santa Rita (38)	LPC 227
	CE: Chapada do Araripe (18)	LPC 246, 290
	MG: Mocambinho (78)	MNRJ 43810 *
P. brasiliensis	MG: Fazenda Santa Cruz (80)	AP 48
P. dasythrix	RS: Bairro Agronomia (137)	AC 628
	RS: Parque Estadual de Itapuã (139)	AC 629
P. lamarum	MG: Estação Ecológica de Acauã (79)	LC 73
P. lundi	MG: Fazenda do Bené (90)	MNRJ 62392
P. mantiqueirensis	MG: Fazenda da Onça (91)	MNRJ 62393
P. nigrispinus	RJ: FS 12, Cachoeiras de Macacu (71)	FS 12-03, 12-30
	RJ: FS 6, Cachoeiras de Macacu (70)	FS 6-43
P. pattoni	BA: Fazenda Monte Castelo (51)	YL 198, 199, 200, 201
	BA: Helvécia (53)	HGB 36
	BA: Mangue do Caritoti (50)	MNRJ 62391; YL 195, 196
	BA: Una (49)	VF 19
	ES: Estação Biológica de Santa Lúcia (57)	MBML 2011, 2032, 2047
	ES: Fazenda Santa Terezinha (54)	LC 32
	MG: Mata Paraíso (85)	UFV 696
	RJ: Fazenda Rosimery (67)	FS 11-52
	RJ: Fazenda São José da Serra (62)	MVZ 183139
	RJ: Fazenda União (68)	MNRJ 42978

* Sequence from Lara et al. (1996).

The maximum parsimony analysis resulted in 32 most parsimonious trees, each 1001 steps long (CI = 0.4611; RI = 0.7718). The differences between these trees are restricted to the terminal branches in one clade comprising 17 individuals of a single species, *P. pattoni*, having little sequence divergence among them. Figure 9 shows the strict consensus tree and support for internal nodes. There is very strong support for the monophyly of *Phyllomys*, with 99% bootstrap value and 14 steps of Bremer support. In the shortest tree, *P. mantiqueirensis* is the sister taxon to all other species of *Phyllomys*, but with very little support (bootstrap < 50%, Bremer = 1). *Phyllomys pattoni* forms the next clade, sister to all remaining species, which are supported by a bootstrap value of 87% and 6 steps of Bremer support. Then, the robust southern clade (*P. dasythrix* and *P.* aff. *dasythrix*) is sister to a weakly supported group formed by the northeastern clade (*P. blainvilii, P. lamarum,* and *P. brasiliensis*) and a clade joining *P. lundi* and *P. nigrispinus*. Within the northeastern clade, *P. brasiliensis* and *P. lamarum* form a monophyletic group sister to *P. blainvilii*.

According to the likelihood ratio test, the model of nucleotide substitution that best fits the data is the General Time Reversible model (Rodríguez et al., 1990), taking into account the proportion of invariable sites and variable sites following a gamma distribution (GTR+I+G). The maximum likelihood tree (Fig. 10) had a score of -Ln = 6007.2, and a topology similar to that of the parsimony tree, with the same well-supported internal clades of the northeast (*P. blainvilii, P. lamarum,* and *P. brasiliensis*, bootstrap = 91%) and south (*P. dasythrix* and *P.* aff. *dasythrix*, bootstrap = 99%). The difference between these two trees comes from weakly supported clades. Figure 11 shows a comparison where two differences in topology can be observed: (I) the placement of *P. pattoni* as sister to all other species in the likelihood tree and of *P. mantiqueirensis* in the parsimony tree; (II) *Phyllomys lundi* and *P. nigrispinus* as sister groups in the parsimony tree and sister to the northeastern and southern clades respectively, in the likelihood tree. The first disagreement is not unexpected, since the support for either *P. mantiqueirensis* or *P. pattoni* as the sister taxon to all other species is extremely low in both analyses (bootstrap < 50%, Bremer = 1). The best way to represent this basal split is as a trichotomy. The placement of *P. lundi* and *P. nigrispinus* as sister taxa in the parsimony analysis, however, had some bootstrap support (69%), and four extra steps are necessary to break their monophyly.

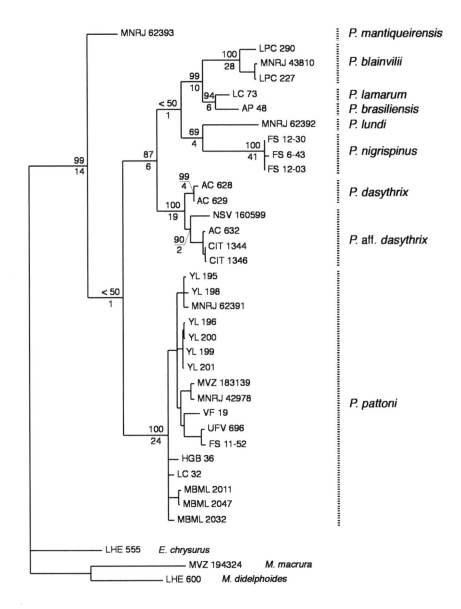

FIGURE 9. Strict consensus of 32 most parsimonious trees (length = 1001 steps each) of *Phyllomys* spp. based on complete sequences of the cyt b gene. Percent bootstrap values are given above branches and Bremer support values are given below. Branch lengths are proportional to the amount of change.

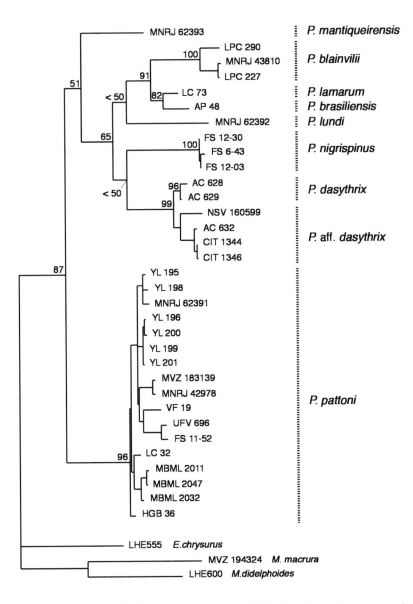

FIGURE 10. Maximum likelihood tree (-Ln = 6007.2) of *Phyllomys* spp. based on complete sequences of the cyt b gene. Percent bootstrap values are given at all major internal nodes.

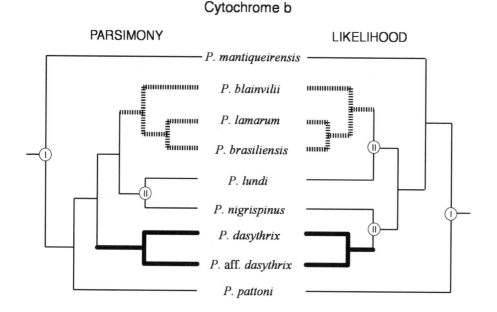

FIGURE 11. Comparison between topologies of the parsimony (left) and likelihood (right) trees of *Phyllomys* spp. The northeastern (hatched lines) and southern (bold lines) clades are well supported in both analyses. Differences in topology involve the placement of *P. pattoni* and *P. mantiqueirensis* (I) and the position of *P. lundi* and *P. nigrispinus* (II).

GENETIC STRUCTURE

Pairwise sequence divergences of the cyt b calculated using the Kimura 2-parameter model are presented in Table 2. Mean interspecific genetic distances vary from 2.7% between *P. dasythrix* and *P. aff. dasythrix* to 13.4% between *P. lundi* and *P. pattoni*, but in most cases values are above 10% (Fig. 12). The average genetic divergence within species is usually around 2.0%, and the absolute maximum is 3.7% between two samples of *P. pattoni* (Fig. 12). This intraspecific pattern is congruent with the geographic location of the samples, which tend to be isolated by distance, as shown in figure 13 for *P. aff. dasythrix* and *P. blainvilii*. In *P. pattoni* the same trend is present, but two elements discussed below are evident: (1) several pairs show a higher divergence than expected given their geographic proximity, and (2) there are no pairs which are closely related but geographically distant (Fig. 13). A haplotype network using a layout that attempts to match the geographic origin of the samples (Fig. 14) allows further understanding of this pattern for *P. pattoni*. Each specimen has a unique haplotype and haplotypes are usually as highly differentiated within localities as they are between localities, with no clear geographic structure, although a north to south trend is apparent. The shape of the species range can explain the absence of haplotypes that are genetically close but geographically distant. Today *P. pattoni* occurs from southeastern to northeastern Brazil, mainly with a narrow range along the coast that gets even narrower towards the northeast. Assuming that this configuration did not change significantly over time, one would expect the haplotype network to follow a relatively linear pattern from north to south, even in the absence of geographic structure since gene flow is likely to occur in a stepping-stone pattern (Endler, 1977).

The geographic structure of cyt b haplotypes of *P. pattoni* seems to be a consequence of large effective population size with no major discontinuities, where mutations are accumulating over time and polymorphisms are maintained. The large number of steps between haplotypes in the network (see Fig. 14) suggests a deep age, which argues against recent expansion. Tests for a possible recent population expansion provided mixed results. The Tajima's D value of -1.03 is negative, therefore indicating the possibility of expansion, but is not significantly different from zero ($p = 0.16$), which represents neutrality. The results of the mismatch distribution (Fig. 15) show a tendency to a polymodal curve, characteristic of samples drawn from populations under demographic equilibrium. However, the mismatch distribution is not significantly different from the distribution of the model curve expected under expansion (SSD = 0.0095, $p = 0.57$; HRi = 0.0231, $p = 0.28$). In conclusion, there are no clear imprints of recent expansion and the sample size may simply be too small to confirm expansion or stability.

TABLE 2. Average percent pairwise sequence divergence between and, when available, within species of the genus *Phyllomys* based on the Kimura 2-parameter model. Outgroups *Echimys chrysurus* and *Makalata* spp. were included for comparison.

	Ppatto (n=17)	Pafdas (n=4)	Pdasyt (n=2)	Pnigris (n=3)	Plundi (n=1)	Pbrasi (n=1)	Plamar (n=1)	Pblain (n=3)	Pmanti (n=1)	Echi
Maka	19.4	19.4	19.1	19.2	19.6	18.7	18.5	18.5	18.4	19.5
Echi	15.6	18.1	17.2	17.7	18.6	15.8	16.0	17.7	14.6	–
Pmanti	10.5	11.8	11.2	12.1	12.4	11.0	10.4	11.4	–	
Pblain	13.3	12.1	11.6	13.0	13.0	7.6	7.1	2.0		
Plamar	11.9	9.9	9.3	10.4	10.0	3.3	–			
Pbrasi	10.4	10.9	9.8	10.6	10.2	–				
Plundi	13.4	12.9	11.7	10.8	–					
Pnigris	12.8	11.2	10.4	0.3						
Pdasyt	11.1	2.7	0.5							
Pafdas	11.9	1.8								
Ppatto	2.0									

Taxon abbreviations: Maka = *Makalata* spp.; Echi = *Echimys chrysurus*; Pmanti = *P. mantiqueirensis*; P. blain = *P. bainvilii*, Plamar = *P. lamarum*; P. brasil = *P. brasiliensis*; Plundi = *P. lundi*; Pnigris = *P. nigrispinus*; Pdasyt = *P. dasythrix*; Pafdas = *P.* aff. *dasythrix*; Ppatto = *P. pattoni*.

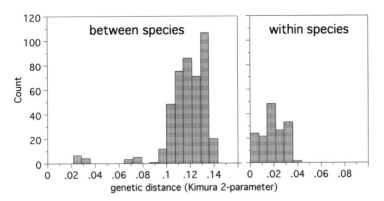

FIGURE 12. Histograms comparing the pairwise genetic distances between (left) and within (right) species of *Phyllomys* using the Kimura 2-parameter model.

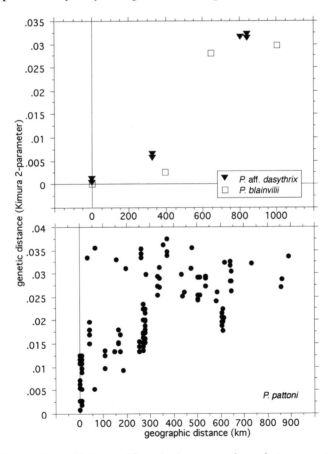

FIGURE 13. Scatterplots of intraspecific pairwise comparisons between geographic versus genetic distances of *P. blainvilii*, *P.* aff. *dasythrix* (above), and *P. pattoni* (below).

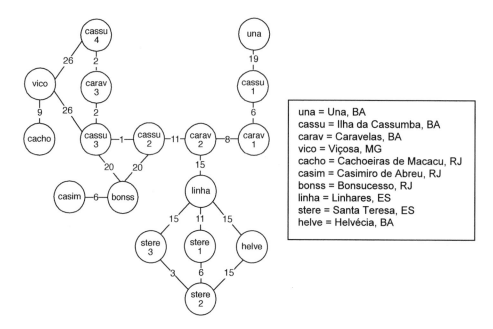

FIGURE 14. Minimum spanning network for haplotypes of *P. pattoni* in a layout that attempts to match geography. The number of bases that differ between pairs of haplotypes that are connected in the network is indicated on the lines connecting them.

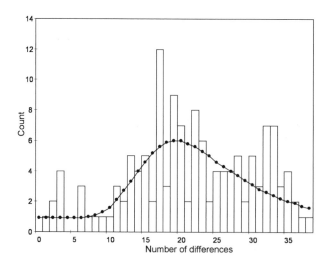

FIGURE 15. Histogram of the number of base differences between all cyt b haplotypes of *P. pattoni* compared with the expected distribution under a model of uniform population expansion (solid line and dots).

BIOGEOGRAPHY

Phylogeographic Patterns

It is possible to reconstruct some events in the biogeographic history of *Phyllomys* using the phylogenetic results based on cyt b sequences. Unfortunately the molecular clock was rejected when likelihood ratio tests were applied to the relationships within *Phyllomys*, therefore preventing date estimates. However, assuming the split between *Phyllomys* and *Echimys* to have happened at about 3.5 million years ago (see Leite and Patton, 2002), the events discussed in this section probably took place during the Late Pliocene and Pleistocene. Within *Phyllomys*, the oldest episode is the nearly simultaneous separation between a lowland coastal clade (*P. pattoni*), the high-elevation *P. mantiqueirensis*, and the common ancestor of the remaining forms (Fig. 16), which probably took place in the Late Pliocene. The next split has a north-to-south component and involves the northeastern clade, the southern clade, and two species in the southeast (*P. nigrispinus* and *P. lundi*). Nevertheless, this event probably took place shortly after the first one, still in the Pliocene, since the levels of sequence divergence are high and quite similar (11.7% and 11.0%).

Vivo (1997) reviewed the mammalian evidence of historical vegetation changes in the semiarid Caatinga of northeastern Brazil based on disjunct distributions, forest-adapted fossil primates found by Hartwig and Cartelle (1996), and scarcity of physiological adaptations to arid climate. The presence of *P. blainvilii* in the brejos of the Caatinga is another testimony of a former contiguous forest. Three main inferences can be drawn from the *P. blainvilii* data:

- The former widespread forest had closer connections to the Atlantic forest than to the Amazon, at least in recent times (Late Pliocene, Pleistocene). This is supported by the fact that *Phyllomys* is the only tree rat found today in the areas of brejos. The easternmost record for the Amazonian genus *Echimys* is on the eastern edge of Amazonia in the state of Maranhão (Oliveira and Mesquita, 1998);
- The isolation has a north-to-south component, since the northernmost forest patch at the Chapada do Araripe separated first from the Andaraí and Mocambinho areas, suggesting that aridity increased progressively from north to south (see Fig. 16);
- The isolation of the forest patches has not increased the levels of genetic divergence within *P. blainvilii*, because it shows the same pattern of divergence and isolation by distance as expected and observed in other species occurring in continuous forest (*P. pattoni* and *P.* aff. *dasythrix*, see Fig. 13). Recent connections between the forest patches could have allowed gene flow among populations of *P. blainvilii*. For example, a connection

between 15,500 and 11,800 years ago due to wetter climatic conditions is
supported by pollen analysis of marine sediment (Behling et al., 2000).

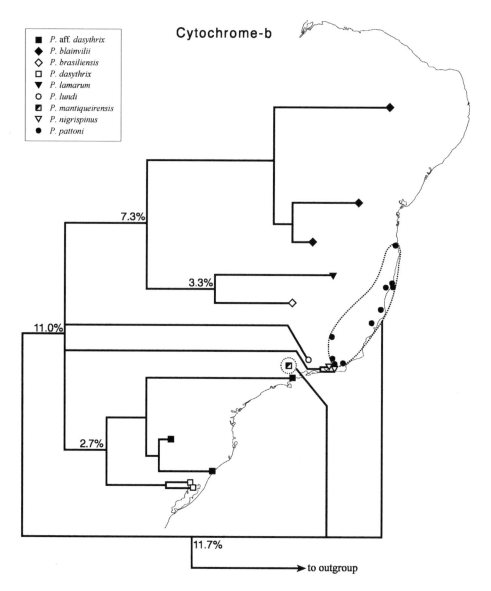

FIGURE 16. Phylogeographic relationships of *Phyllomys* spp. based on complete cyt b
sequences. The cladogram is a strict consensus of the parsimony and likelihood results.
Numbers at the nodes are average pairwise Kimura 2-parameter distances.

The divergence leading to the sister species *P. brasiliensis* and *P. lamarum* is likely very recent (Pleistocene) since they are very close genetically (3.3%) and very similar in general morphology, differing primarily in body size, the former being 15% smaller in head and body length than the latter (see Cranial Morphometrics, below). Today, they are separated by the Serra do Espinhaço, which could have acted as a vicariant barrier especially during recent climatic fluctuations. Nevertheless, considering that the range of *P. brasiliensis* is much more restricted than that of *P. lamarum* and we do not know whether the Espinhaço has been an effective barrier to gene flow, speciation via the formation of a peripheral isolate cannot be discarded (see Lynch, 1989).

Speciation within the southern clade is also very recent and probably occurred during the Pleistocene, since these species diverge by only 2.7%. The two species are virtually identical in cranial morphology and very similar in body size, but differ markedly in pelage stiffness and diploid number: 2n = 72 in the soft-furred *P. dasythrix* and 2n = 92 in the stiff-bristled *P.* aff. *dasythrix* (chromosome data from A. Christoff., pers. comm.). The geographic distribution of the two species is still tentative since there are very few museum records, but there seems to be some overlap in their range. However, no cases of sympatry or hybrids have been detected, and the closest known localities are approximately 75 km apart. The rate of speciation may be correlated with chromosomal change (Bush et al., 1977), which could be involved in the evolution of *P. dasythrix* and *P.* aff. *dasythrix.* For example, a modified form of stasipatric speciation (White, 1968) where the chromosomal rearrangement establishes itself in a peripheral colony and then spreads through the existing species population is a possible scenario. The low level of genetic divergence lends support to this hypothesis, but these ideas still need to be confirmed experimentally (see Reiseberg, 2001).

Spatial and Temporal Context

Haffer (1969) and Vanzolini (1970) independently proposed the refuge theory for tropical America, based on concordant patterns of rainfall and endemism of birds and lizards, respectively. They suggested that climatic changes and the cyclic formation of isolated rainforest patches (refugia) were the cause of the patterns of distribution and diversity we observe today in lowland Amazonia. Similar biogeographic patterns were recorded later for plants, butterflies, and other organisms, and they were correlated with geomorphological data and paleoenvironmental reconstructions of the Pleistocene (references in Prance, 1982). These data served as the evidence for the refuge theory and the concept of Neotropical refuge biogeography, which has been widely accepted and used as a framework for several applications, including conservation and land use planning (Whitmore and Prance, 1987).

Debates over the refuge theory reintensified over the last decade as new paleopalynological data became available. Although there is a consensus about cooling, aridity has been strongly rejected by Colinvaux and collaborators (Bush et al., 1990; Colinvaux et al., 1996a; Colinvaux et al., 1996b), who argue that tropical forests persisted through recent glacial cycles, but plant communities were not necessarily the same (for reviews, see Colinvaux et al., 2000; Colinvaux and Oliveira, 2001). In addition, regional patterns differ from the general model of moisture changes during glacial periods (Auler and Smart, 2001; Salgado-Labouriau et al., 1998). The eastern Brazilian forests have been treated only marginally in the refuge debate, although a chain of several small isolated refuges (areas with 80–100% likelihood of persistence) have been proposed that could actually have been connected (through areas with 60% likelihood of persistence; see summary map in Brown, 1987). The validity of the refuge hypothesis cannot be addressed with *Phyllomys* or any phylogenetic data, because the biogeographic process itself is not testable (Lynch, 1988). Testing the Pleistocene age of speciation through biological clocks as suggested by Lynch (1988) does not falsify the modern version of the refuge theory, since it now postulates refuges during most of earth's history, following Milankovitch cycles (Haffer, 1993). I therefore address the tectonic, climatic, and vegetation changes in eastern Brazil since the Pliocene that set the stage for the diversification of the Atlantic tree rats.

An important component in this context is the history of the seasonally dry forests, usually neglected in discussions of vegetation changes in the Quaternary (Pennington et al., 2000). Based on disjunct distributions, Prado and Gibbs (1993) argued that dry forests that are isolated today were once an extensive and largely contiguous dry woodland formation. Pennington et al. (2000) suggested that a form of seasonally dry tropical forest, adapted to cooler conditions instead of open formations as traditionally postulated by the refuge theory, replaced some of the humid forests during climatic changes. This hypothesis, if true, could have played an important role in the evolution of the tree rats since some species of the northeastern clade are now restricted to semideciduous forests inland, while others occur mainly in the coastal rainforests (e.g., *Phyllomys pattoni*). Lara and Patton (2000) observed a similar pattern in the Atlantic spiny rats of the genus *Trinomys*, since two main clades are restricted to either dry interior or moist coastal forests. They hypothesized that the differentiation and current distribution within this group is associated with the uplift of the coastal mountains generating a rain-shadow effect and consequently the current vegetation-humidity zones.

The complex topography and associated tectonic activities in southeastern Brazil are a crucial element in the context of landscape evolution and consequently of the fauna and flora. The Brazilian Highlands (terrains rising ca. 500 m above sea level) are uplifted areas of Precambrian and Paleozoic crystalline and sedimentary formations (Petri and Fúlfaro, 1983). During the Cenozoic, tectonic activities were intense in the eastern portion of southeastern Brazil, resulting in very steep hills.

These movements are associated with a fault system extending in a southwest-northeast direction, parallel to the Atlantic coast, most evident between 22° and 28° latitude south (Almeida, 1976; Clapperton, 1993; Petri and Fúlfaro, 1983). The topographic evidence of this system is the impressive cliffed edges of escarpments, like the Serra da Mantiqueira, the mountain range that runs parallel to the Atlantic coast inland to the Serra do Mar, today rising 2000 m above sea level. Although there is still seismic activity and uplift today, most of the uplift took place during the Pliocene and Quaternary (Almeida, 1976). Unfortunately there are few data available documenting either the precise dates of these events or the rate of uplift. According to Almeida's (1976) reconstruction, a tectonic pulsation, probably during the Pliocene, affected the Serra do Mar system, and occurred with an accentuation of the relief of mountains and rifts, which at that time assumed their modern aspect (see also Petri and Fúlfaro, 1983). The tectonic activities and uplift of the coastal mountains must have had profound effects on the distribution of the vegetation, especially moist and deciduous forests.

Glaciation probably never occurred in South America outside the Andes, although geocryogenic activity induced by periodic freezing and thawing of ground ice took place in areas above 2000 m in eastern Brazil, especially in the Itatiaia massif, Serra da Mantiqueira (Clapperton, 1993). This resulted in a landscape dominated by stripped bedrock surfaces, boulder streams, and geocryogenic 'head' deposits. Today the summits are usually covered with high-altitude grassland, or *campos de altitude* (Safford, 1999a, b) above elevations of 1800–2000 m. Many plant taxa in these *campos de altitude* show phytogeographic affinities with the *páramos* in the Andes, confirming the past connections of these floras through southern Brazil during periods of favorable (cooler) climate beginning in the early Cenozoic (Safford, 1999a), and therefore displacing high-elevation forests.

In extra-Andean South America, topography achieves maximum complexity in southeastern Brazil, where several species of *Phyllomys* occur either in parapatry or in limited sympatry (Fig. 17). This region also marks the break between the northeastern and southern clades. Vanzolini (1988) noted very similar patterns in Atlantic forest lizards: a distributional break at about 19°S and the highest diversity at 21–22°S. He highlighted the importance of latitude and hypothesized that temperature played a major role in defining this distribution pattern. A similar pattern is found in the Atlantic tree rats as well; unrelated species are distributed either north or south of 20°S and the highest diversity (8 species) is found between 22° and 24°S. This pattern could also be the result of temperature zonation in terms of both latitude and altitude. Costa et al. (2000) found the same general trend in a parsimony analysis of endemicity using raw distributions of mammals. According to these authors, areas to the north and to the south of the Serra da Mantiqueira are

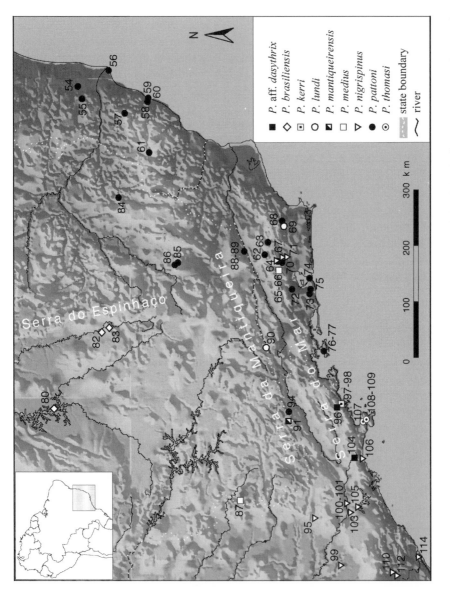

FIGURE 17. Map of southeastern Brazil showing collecting localities of *Phyllomys* spp. superimposed on a digital elevation model. Numbers correspond to numbered localities in the Gazetteer.

reciprocally monophyletic and the highest species diversity in the Atlantic forest was found between 21° and 24°S.

In conclusion, the distribution and diversity we observe in the Atlantic tree rats today are the result of both deep and shallow historical events, namely the dynamic interplay of tectonic, climatic, and vegetational changes since the Late Tertiary associated with the natural history and ecological requirements of the species. Unfortunately the data available do not allow us to establish precisely what events were the most important or when they occurred.

THE GENUS

Phyllomys Lund

Type species: *Nelomys blainvilii* Jourdan, 1837

Distribution and Habitat

Species of *Phyllomys* are found in forested areas along eastern Brazil, from the states of Paraíba and Ceará (ca. 7°S) in the northeast to Rio Grande do Sul (ca. 30°S) in the south, reaching the São Francisco and Paraná river basins to the west (approx. 54°W). They occur from sea level along the coast up to 1850 m of elevation in the Serra da Mantiqueira, in rainforests and associated ecosystems such as mangroves and semideciduous forests of the interior, including the "brejos" in the Brazilian Caatinga. They have not been recorded in neighboring countries with contiguous forests, such as Argentina (Cabrera, 1961; U. Pardiñas, pers. comm.) or Paraguay (Myers, 1982).

Diagnosis

Members of the genus *Phyllomys* are arboreal rodents with spiny to soft fur, large eyes, small and round ears, and long whiskers reaching the shoulders. The limbs are short, fore and hindfeet are broad and short, with strong claws on all digits except the thumb, which bears a nail. Dorsal surface is brown to reddish golden brown, and venter ranges from pure white to light gray-brown. Spines are conspicuous in most species, especially on the rump, where they can reach 1.5 mm in width in some species. A well-developed sternal gland is present on the chest of individuals of both sexes, but more evident in males. Tail is slightly shorter to longer than head and body. Tail base is covered by body-like hair for approximately 20 mm, the remaining length varying from almost naked to thickly haired with a bushy tuft at tip. Skull is strong, with relatively short rostrum, bulla is usually inflated, interorbital region is wide, palate is narrow and toothrows are nearly parallel, incisive foramina are small to relatively large. Upper molars are rectangular and composed of four transverse laminae separated by three labial flexi (para-, meso- and metaflexus). In young and adult individuals, the enamel of one lamina does not contact the others, but in old animals, two or more laminae coalesce at the labial and/or the lingual side, forming a transverse U-shaped loph, similar to that in *Echimys* and *Makalata*. The pattern of coalescence varies (Moojen, 1952; Winge, 1887), but usually the two posterior lophs unite labially with wear, and the anterior lophs unite lingually (Emmons, in review). In the lower dentition, the hypoflexid is deep and angled forward and connects the mesoflexid on M1, but a weak mure may form between the two flexi with age. A slender mure is typically present on the second and third molars, separating the hypoflexid and mesoflexid, giving it the shape of the Arabic numeral 3. This mure is usually more developed on M2, and especially M3. The lower premolar is pentalophodont, the first two lophids coalesce at an early age forming a triangle with a circular island of enamel inside. The metaflexid bisects the tooth only in young individuals, but the mesoflexid connects the hypoflexid even in adults.

Emmons (in review) provided additional characters that diagnose the genus *Phyllomys*, including the following: four pairs of mammae, three lateral and one inguinal; auditory tympanic bullae with auditory meatus low, directed slightly forward, with a space as wide as the meatus between it and the squamosal; additional rings of bone external to meatus are often present.

Comparisons with *Echimys*, *Makalata*, and *Diplomys*

Given the taxonomic confusion regarding scientific names applied to tree rats, in the following comparisons, I use the exemplar species *Echimys chrysurus* (genotype of *Echimys*), *P. blainvilii* (genotype of *Phyllomys*), *Makalata macrura*, and *Diplomys labilis* when referring to each of these genera, unless otherwise specified. Morphological distinctions between these four genera include tail length, usually shorter than head and body in *Diplomys* and *Makalata* and longer in *Echimys* and *Phyllomys*. Aristiforms are heavily spiny in *Echimys* and *Makalata*, soft in *Diplomys*, and soft to heavily spiny in *Phyllomys*, depending on the species (Fig. 18). The orbits are larger, the interorbital region is narrower, the nasals are shorter, and the zygomatic arch is more slender in *Diplomys* as its ventral edge does not expand posteriorly forming a process as in *Echimys*, *Makalata*, and *Phyllomys* (Fig. 19). The incisive foramen is typically longer in *Phyllomys* than in the other three genera, and the mesopterygoid fossa is wider in *Echimys* and *Makalata* than in the other two genera.

As pointed out by several authors (e.g., Emmons, in review; Moojen, 1952; Thomas, 1916a), the main distinction between *Phyllomys* and other echimyine genera is the dentition (Fig. 20). In *Phyllomys* the upper cheekteeth consist of four independent transverse laminae, as in *Diplomys*. In the upper molars of *Echimys*

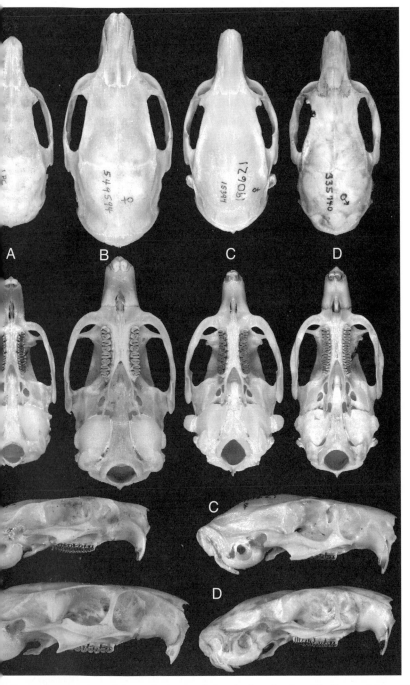

Dorsal, ventral, and lateral views of the skulls of: A—*Phyllomys blainvilii*
B—*Echimys chrysurus* (USNM 549594); C—*Makalata macrura* (MVZ
D—*Diplomys labilis* (USNM 335740). Natural size.

and *Makalata*, however, the anteroloph and protolop
mesoloph and metaloph, each pair forming a transv
be connected in the lower dentition of *Phyllomys* fo
"3" similar to *Echimys* and *Makalata*, but the hy
strongly angled in *Phyllomys*. In *Diplomys,* the lop
independent laminae.

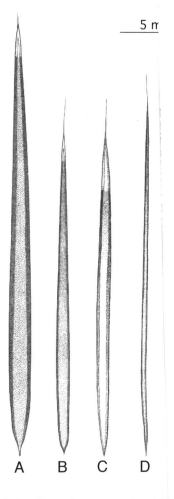

5 m

A   B   C   D

FIGURE 18.   Aristiform hairs from the rump of echin
(USNM 549594); B—*Makalata macrura* (MVZ 190621);
290); D—*P. dasythrix* (MCNFZB 46); E—*Diplomys labil*
pictured in white is actually orange/ochraceous in A, B, and

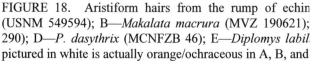

FIGURE
(LPC 2
190621);

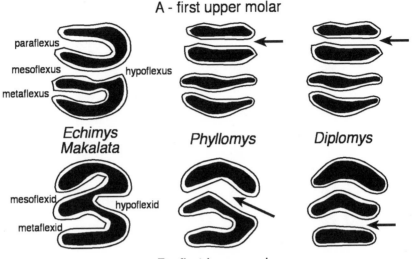

FIGURE 20. Diagrams comparing echimyine tooth morphology. A—upper M1: paraflexus does not bisect tooth in *Echimys* and *Makalata* (left), paraflexus bisects tooth in both *Phyllomys* (middle) and *Diplomys* (right). B—lower M1: mesoflexid does not connect hypoflexid and metaflexid does not bisect tooth in *Echimys* and *Makalata* (left), mesoflexid connects hypoflexid in *Phyllomys* (middle), mesoflexid connects hypoflexid and metaflexid bisects tooth in *Diplomys* (right).

Thomas (1916a, p. 295) clearly pointed out that in *Phyllomys* "the upper molars consist of four simple transverse laminae, which persist without coalescing with each other at different ages, this continued separation being due to the equal depth of the three transverse valleys dividing the laminae" (see Fig. 2). Since then, most authors ignored his observation, and I believe the historical reason behind lumping *Phyllomys* within *Echimys* is the very small number of museum specimens examined by scholars. For instance, Tate (1935, p. 431) stated "I have not seen a number of the east Brazilian species" but he separated "*Echimys*" (including *Phyllomys*) into a hairy-tailed group and a scaly-tailed group, both paraphyletic. Ellerman (1940, p. 109) also pointed out that "not very much material of this interesting genus [*Echimys* including *Phyllomys*] is available for examination," and he decided not to use the name *Nelomys* or *Phyllomys* for the southern Brazilian types. Indeed, with very few museum specimens, one would have difficulty separating these two genera, especially if very old individuals of *Phyllomys* with coalesced lophs are included in the sample. In *Echimys*, however, the lophs are coalesced at the time of tooth eruption (Fig. 21A), while in *Phyllomys*, they are independent at the time of eruption (Fig. 21B) and only come into contact after substantial wear (see Fig. 2).

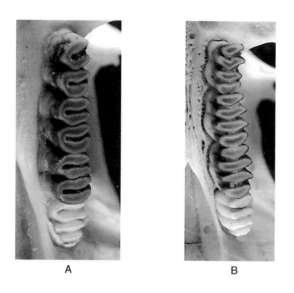

A   B

FIGURE 21. Left maxillary toothrows of: A—*Echimys chrysurus* (USNM 549595); and B—*Phyllomys blainvilii* (MNRJ 43810). Note the coalesced lophs on the occlusal surface of erupting M3 in *Echimys* but not in *Phyllomys*.

Since the degree of differentiation in tooth morphology is the character traditionally used to diagnose echimyid genera (Ellerman, 1940; Thomas, 1916a, b), and considering that *Phyllomys* is a monophyletic group readily diagnosable by a unique combination of dental characters, there is no reason for lumping it with *Echimys*. Moreover, *Echimys* as it is known in the recent systematic literature (e.g., Woods, 1993) probably comprises a paraphyletic cluster of tree rats (Carvalho, 1999; Emmons, in review) even when *Phyllomys* is excluded from it.

Local Names

The use of common names varies widely in Brazil according to geography. The most common local name in Portuguese associated with species of *Phyllomys* is *rato de espinho* (spiny rat). The only other echimyid with spiny fur co-occurring throughout the range of *Phyllomys* is the terrestrial *Trinomys*. Usually, when these occur together, locals differentiate them, calling the latter *barriga-branca*, referring to the pure white belly of *Trinomys*. The following names have also been used locally to identify species of Atlantic tree rat: *P. blainvilii*: *rabudo vermelho* (Bom Jesus da Lapa, BA), *rato-coandu* (Viçosa, AL), *rato-corói* (Garanhuns, PE); *P. thomasi*: *cururuá*, *cururuá sem rabo* (Ilha de São Sebastião, SP).

SPECIES ACCOUNTS

In this section, I summarize characters that will aid workers in identifying species of *Phyllomys* in the field and in museum collections. Figure 22 illustrates a few skull characters, and table 3 summarizes diagnostic features. Figure 5 shows the differences in aristiform size and stoutness, and figures 23 and 24 depict the skulls of all species side by side for comparison. I list type specimens and type localities following Emmons et al. (2002), known geographic distribution, and specimens examined for each taxon. The distribution maps presented are tentative, because most species are known from very few localities. The known localities were simply connected with an approximate minimum convex polygon. Emmons et al. (2002) present a taxonomic review based on type specimens, including lists of synonyms and restriction of type localities.

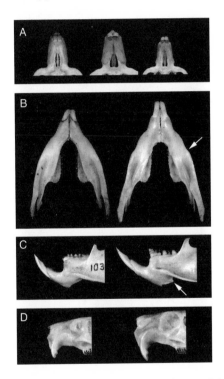

FIGURE 22. Selected diagnostic skull characters. A—Shape of the incisive foramen (from left): ovate, teardrop, bullet. B—Anterior-most ventral origin of the angular process: not deflected laterally (left), deflected laterally (right). C—Ventral mandibular spine: absent (left), present (right). D—Angle of upper incisors: orthodont (left), opisthodont (right).

TABLE 3. Selected diagnostic characters of the species of *Phyllomys*. EAM = external auditory meatus.

Character	P. blaimvilii	P. brasiliensis	P. dasythrix	P. aff. dasythrix	P. kerri	P. lamarum	P. lundi
Body size (greatest skull length)[1]	small-medium (44.4 – 52.7 mm)	medium (48.5 – 52.3 mm)	small-medium (46.0 – 52.0 mm)	samll-medium (47.1 – 50.7 mm)	medium (51.7 mm)	small-medium (43.1 – 51.3 mm)	small (47.7 mm)
Texture of dorsal pelage	spiny	spiny	soft	spiny	spiny	spiny	spiny
Size of aristiforms on rump	medium length and width (24 x 1.0 mm)	medium length, wide (27 x 1.3 mm)	medium length, thin (26 x 0.2 mm)	medium length and width (26 x 0.6 mm)	medium length and width (27 x 1 mm)	medium length, wide (24 x 1.3 mm)	medium length and width (22 x 0.8 mm)
Aristiform tip	whip-like, orange	whip-like, orange	whip-like, black	whip-like, black	whip-like, black	whip-like, orange	whip-like, orange
Tail tip	tufted	hairy	hairy	hairy	hairy	nearly naked	hairy
Palatine width at M1	> tooth width	≈ tooth width	< tooth width	< tooth width	≥ tooth width	≤ tooth width	≈ tooth width
Maxillary toothrows	slightly divergent at either end	parallel	parallel	parallel	parallel	slightly divergent at either end	slightly divergent posteriorly
Supraorbital ridges	well developed	well developed, beaded	weakly developed	weakly developed	well developed	well developed	weakly developed
Interorbital region	divergent posteriorly	divergent posteriorly	divergent posteriorly	divergent posteriorly	divergent posteriorly	divergent posteriorly	divergent posteriorly
Postorbital process	absent or inconspicuous	present	inconspicuous	inconspicuous	inconspicuous	absent or inconspicuous	absent
Zygomatic arch height	≤ 1/3 jugal length	≤ 1/3 jugal length	≈ 1/3 jugal length	≈ 1/3 jugal length	≈ 1/3 jugal length	≤ 1/3 jugal length	≈ 1/3 jugal length
Postorbital process of zygoma	usually spinose, formed mainly by jugal	spinose, formed by jugal only	rounded or spinose, formed mainly by jugal	spinose, formed mainly by jugal	spinose, formed mainly by jugal	spinose, usually formed by jugal only	spinose, formed by jugal and squamosal
Mastoid process	short, extending to midline of EAM	long, extending below midline of EAM	short, extending to midline of EAM	short, extending to midline of EAM	short, extending to midline of EAM	short, extending to midline of EAM	short, extending to midline of EAM
Incisive foramen	ovate	ovate	ovate	ovate	ovate	ovate	ovate
Ventral root of the angular process	laterally deflected	laterally deflected	does not deflect laterally	does not deflect laterally	does not deflect laterally	laterally deflected	laterally deflected
Ventral mandibular spine	absent	present	absent	absent	absent	present	absent
Upper incisors	orthodont	orthodont	orthodont	orthodont	slightly opisthodont	orthodont	orthodont

TABLE 3 (continued)

Character	P. mantiqueirensis	P. medius	P. nigrispinus	P. pattoni	P. thomasi	P. unicolor
Body size (greatest skull length)[1]	small (48.1 mm)	medium (46.2 – 56.7 mm)	medium (45.6 – 55.5 mm)	medium (44.7 – 54.6 mm)	large (56.2 – 65.7 mm)	large (57.5 mm)[2]
Texture of dorsal pelage	soft	stiff	spiny	spiny	spiny	stiff
Aristiforms on rump	medium length, thin (24 x 0.1 mm)	very long, thin (36 x 0.4 mm)	medium length and width (27 x 1.0 mm)	medium length and width (23 x 1.0 mm)	long, medium width (33 x 0.7 mm)	short, thin (20 x ? mm)
Aristiform tip	whip-like, black	whip-like, black	whip-like, black	blunt, orange	whip-like, black	whip-like, black
Tail tip	tufted	hairy	hairy	nearly naked	hairy	?
Palatine width at M1	< tooth width	≈ tooth width	≥ tooth width	≈ tooth width	≥ tooth width	< tooth width
Maxillary toothrows	slightly divergent posteriorly	parallel	parallel	slightly divergent posteriorly	parallel	parallel
Supraorbital ridges	weakly developed	well developed	well developed	well developed, beaded	well developed	well developed
Interorbital region	divergent posteriorly	divergent posteriorly or parallel sided	divergent posteriorly	divergent posteriorly	parallel sided	?
Postorbital process	absent	absent or inconspicuous	absent or inconspicuous	absent	inconspicuous	?
Zygomatic arch height	≈ 1/3 jugal length	> 1/3 jugal length	≤ 1/3 jugal length	≈ 1/3 jugal length	> 1/3 jugal length	≈ 1/3 jugal length
Postorbital process of zygoma	spinose, formed mainly by squamosal	spinose, formed mainly by jugal	rounded or spinose, formed mainly by jugal	spinose, formed by jugal and squamosal	usually rounded, formed mainly by jugal	rounded, formed by jugal only
Mastoid process	long, extending below midline of EAM	short, extending to midline of EAM	long, extending below midline of EAM	long, extending below midline of EAM	short, extending to midline of EAM	long, extending below midline of EAM
Incisive foramen	ovate	tear-drop shaped	ovate	bullet shaped	tear-drop shaped	ovate
Ventral root of the angular process	laterally deflected	does not deflect laterally	does not deflect laterally	laterally deflected	does not deflect laterally	laterally deflected
Ventral mandibular spine	absent	absent	absent	absent	absent	absent
Upper incisors	orthodont	opisthodont	slightly opisthodont	orthodont	slightly opisthodont	orthodont

[1] Values given are minimum and maximum. [2] Skull of the holotype and only known specimen of *P. unicolor* is broken. Given that basilar length (BaL) is highly correlated with GSL ($R^2 = 0.945$), the GSL was estimated through the regression equation GSL = 4.757 + 1.174 * BaL, where BaL = 44.91 mm.

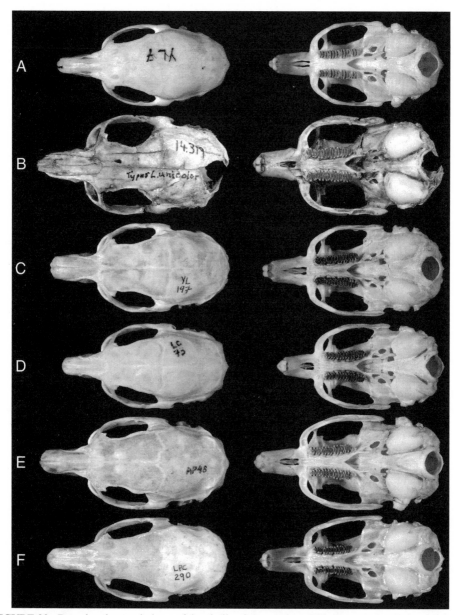

FIGURE 23. Dorsal and ventral views of the skulls of six species of *Phyllomys*: A—*P. lundi* (MNRJ 62392, the holotype); B—*P. unicolor* (SMF 4319, the holotype); C—*P. pattoni* (MNRJ 62391, the holotype); D—*P. lamarum* (LC 73); E—*P. brasiliensis* (AP 48); and F—*P. blainvilii* (LPC 290). Natural size.

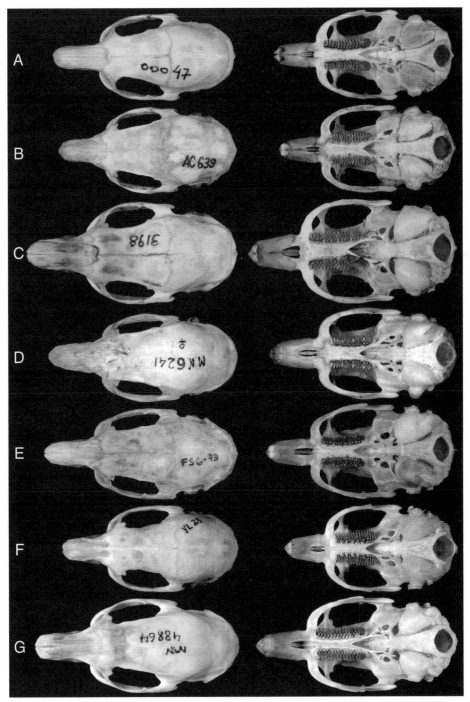

FIGURE 24. Dorsal and ventral views of the skulls of seven species of *Phyllomys*: A—*P. dasythrix* (MCNFZB 47); B—*P.* aff. *dasythrix* (AC 639); C—*P. thomasi* (MZUSP 3198); D—*P. kerri* (MNRJ 6241, the holotype); E—*P. nigrispinus* (FS 6-43); F—*P. mantiqueirensis* (MNRJ 62393, the holotype); and G—*P. medius* (MNRJ 48864). Natural size.

Phyllomys blainvilii (Jourdan, 1837)

Type material: The lectotype MHNG 250/19 is a skin with included skull designated by Emmons et al. (2002), and paralectotypes include MNHG 324/1 and BMNH 55.12.24.116. Specimens were collected by J. S. Blanchet in "Bahia, Brazil" and the type locality was restricted to Seabra, Bahia, Brazil, ca. 12°25'S 41°46'W, by Emmons et al. (2002).

Diagnosis: A small- to medium-sized species. Dorsal surface spiny, ochraceous-brown streaked with black, paler on sides. Aristiforms average 24 mm long and 1 mm wide, pale at the base, darker in the middle, and orange near the thin, whip-like tip (Fig. 5C). Venter pale cream with a hint of yellowish overwash. Tail robust, usually slightly longer than head and body, thickly covered with long blackish brown or gold hair forming a 15 mm, slightly wavy tuft at the tip. Tail darkens from base to tip, usually strongly contrasting with pale color of body. Cheekteeth narrow, palatine width larger than tooth width at M1, toothrows slightly divergent at either end (Fig. 23F). Supraorbital ridges well developed. Interorbital region nearly straight, diverging posteriorly. Zygomatic arch slender, maximum height less than or equal to 1/3 of jugal length. Postorbital process of zygoma usually spinose and formed mainly by the jugal. Mastoid process short, extending to the horizontal midline of the external auditory meatus. Mesopterygoid fossa reaches the anterior lamina of M3; narrow, forming an angle of about 45 degrees. Incisive foramen ovate. Ventral root of the angular process of the mandible deflected laterally. Upper incisors orthodont.

Karyotype: Chromosome data are available for one individual (LPC 277, male) from Andaraí, BA (Fig. 1, locality 38). The diploid number is 50 and the karyotype consists of 19 pairs of chromosomes grading in size from large to small (Fig. 25). The first five pairs are submetacentric, the fifth pair has a distinct, near-terminal constriction, and pairs 6 to 18 are metacentric. The quality of the cell suspensions does not allow categorization of the five remaining autosomal pairs. The sex chromosomes are larger than any autosomal pair, the X chromosome is submetacentric and the Y is subtelocentric. The karyotype of a specimen referred to as *P. blainvilii* was described by Souza (1981), who reported 2n = 50, FN = 94 for a male from Igarassu, PE (Fig. 1, locality 22). The general chromosomal morphology is very similar to that I report here, the main difference being the presence of a pair of large acrocentric chromosomes.

Remarks: Specimens of *P. blainvilii* are the most abundant species of *Phyllomys* in museum collections with approximately 200 individuals, but the species is known from few localities. This taxon is sometimes confused with *P.*

lamarum in the literature (Emmons and Feer, 1997). Moojen (1952), for example, had them reversed; his plate 19 actually portrays an adult *P. lamarum*, identifiable by the speckled fur on the back and nearly naked tail, while plate 20 pictures a young *P. blainvilii* with a more uniform body color and a tufted tail.

FIGURE 25. Metaphase cell of *P. blainvilii* (LPC 227), 2n = 50.

Distribution and habitat: *Phyllomys blainvilii* is found inland in northeastern Brazil from the southern portion of the state of Ceará (ca. 7°S) to the northernmost part of Minas Gerais (ca. 15°S, Fig. 26). Although it ranges throughout a large area in northeastern Brazil, its habitat is fragmentary. It is found mainly in isolated areas of semideciduous forest, especially forest islands within the Caatinga ("brejos") and forest patches along the São Francisco River, but occurs on the coast of Alagoas and Pernambuco.

Specimens examined (Total = 191): **Alagoas**: Sítio Angelim, Viçosa (MNRJ 21512, 21513), **Bahia**: 12 km de Lapa, Bom Jesus da Lapa (MNRJ 4137), Bom Jesus da Lapa (MZUSP 6146), Lapa, Bom Jesus da Lapa (MNRJ 4125, 4126, 4127, 4128, 4129, 4130, 4131, 4132, 4133, 4134, 4135, 4136, 4138, 4139, 4140), Rio São Francisco, Bom Jesus da Lapa (MNRJ 52090), Fazenda Santa Rita, 8 km E Andaraí (LPC 227), specific locality unknown [probably Várzea da Canabrava, Seabra] (MNRJ 2P.1641, MNRJ 2P.1647, MNRJ 2P.1664), Várzea da Canabrava, Seabra (MNRJ 21626, 21627, 21628, 21629, 21630, 21631, 21632, 21633, 21634, 21635, 21636, 21637, 21638, 21639, 21640, 21641, 21642, 21643, 21644, 21645,

21646, 21647, 21648, 21649, 21651, 31542, 31543, 31544, 31554), **Ceará**: Chapada do Araripe, 7 km SW Crato (LPC 246, LPC 290), Crato (MNRJ 1350, MZUSP 6147), Sítio Anil, Crato (MNRJ 21552, 21590, 31538, 31539), Sítio Baixa do Maracujá, Crato (MNRJ 21539, 21540, 21542, 21546, 21547, 21548, 21549, 21550, 21584, 21619, 21620, 21621), Sítio Belo Horizonte, Crato (MNRJ 21518, 21520, 21522, 21523, 21554, 21555, 21556, 21557, 21558, 21559, 21560, 21561, 21562, 21563, 21564, 21565, 21566, 21567, 21591, 21593, 21594), Sítio Caiano, Crato (MNRJ 21519, 21521, 21524, 21571, 21588, 21589, 21592, 21623), Sítio Grangeiro, Crato (MNRJ 21568, 21569, 21624, USNM 304580), Sítio Macaúba, Crato (MNRJ 31534), Sítio Minador, Crato (MNRJ 21570) Sítio Quebra Primeira, Crato (MNRJ 21541, 21585), Sítio Santa Rosa, Crato (MNRJ 7819), Sítio Serra Baixa do Maracujá, Crato (MNRJ 21543, 21544, 21545, 21586, 21587, 21622), Sítio Serra Bebida Nova, Crato (MNRJ 21525, 21526, 21527, 21532, 21533, 21534, 21551, 21572, 21573, 21574, 21577, 21578, 21595, 21596, 21597, 21598, 21599, 21600, 21603, 21618, 21625), Sítio Serra da Inga, Crato (MNRJ 31540), Sítio Serra dos Guaribas, Crato (MNRJ 21528, 21529, 21530, 21531, 21575, 21576, 21601, 21602), Sítio Serrinha, Crato (MNRJ 31535, 31536, 31537), Sítio Trindade, Crato (MNRJ 21579, 21604), Sítio Urucu de Fora, Crato (MNRJ 21535, 21536, 21537, 21538, 21580, 21581, 21582, 21583, 21605, 21606, 21607, 21608, 21609, 21610, 21611, 21612, 21613, 21614, 21615, 21616, 21617, 31533), specific locality unknown [probably Crato] (MNRJ 1345, 1512, 1516, 1517, 1521, 1523, 1528, 1548, 1762), specific locality unknown (MNRJ 4258), **Minas Gerais**: Mocambinho, Jaíba, MNRJ 43810, **Pernambuco**: Sítio Cavaquinho, Garanhuns (MNRJ 21514, 21515, 21516), Reserva Ecológica Charles Darwin (AL 3596), **unknown locality** (MNRJ 2238, UFPB APO62).

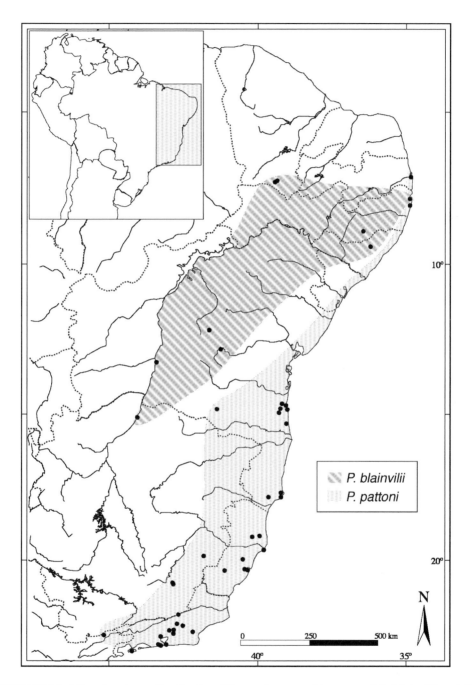

FIGURE 26. Map showing collecting localities and geographic range of *P. blainvilii* and *P. pattoni*.

Phyllomys brasiliensis Lund, 1840

Type material: The lectotype is a maxillary fragment from a young individual figured in Lund (1840a, Plate XXI, Fig. 13) and designated by Emmons et al. (2002). The specimen belongs to the paleontological collection in the Universitets Zoologisk Museum at Copenhagen and it has no catalog number. The type locality is Lapa das Quatro Bocas, near Lagoa Santa, Minas Gerais, Brazil (Emmons et al., 2002).

Diagnosis: A medium-sized species. Dorsal pelage spiny, orange-brown sprinkled with black. Aristiforms orange-tipped, relatively long (27 mm) with a whip-like tip, and wide (1.3 mm) on the rump (Fig. 5A). Venter cream-yellow; inguinal and axillary regions cream-white with white-based hair along the midline. Tail about equal to head and body length; covered with short brown hair; scales visible except near the tip, where longer hairs reach 20 mm. Cheekteeth narrow, palatine width equal to tooth width at M1; toothrows nearly parallel, diverging posteriorly in older individuals. Supraorbital ridges well developed, beaded. Interorbital region slightly divergent posteriorly, postorbital process moderate (Fig. 23E). Zygomatic arch slender to moderately robust, maximum height less than or equal to 1/3 of jugal length. Squamosal does not contribute to the spinose postorbital process of zygoma. Mastoid process long, extending below the horizontal midline of the external auditory meatus. Mesopterygoid fossa sharply pointed, wide posteriorly, forming an angle of more than 60 degrees, but narrower at anterior point, which reaches posterior lamina of M2. Incisive foramen ovate. Ventral root of the angular process of the mandible deflected laterally and ventral spine present posterior to the junction of the mandibular rami. Upper incisors orthodont.

Karyotype: Unknown.

Distribution and habitat: Known only from the Lagoa Santa region, including Sumidouro, and Fazenda Santa Cruz, 150 km to the northwest, all in Minas Gerais state in the valleys of the Paraopeba and das Velhas rivers (Fig. 27). This region is a contact zone between semideciduous forest and Cerrado, sitting between 500 and 800 m above sea level. Parizzi et al. (1998) provided a detailed paleoenvironmental reconstruction of the Holocene of Lagoa Santa.

Remarks: This taxon is known from very few complete museum specimens. Besides several fragmentary skulls from owl pellets collected by Lund (Lund, 1840a, b) and described in more detail by Winge (1887), there are only three adult specimens preserved, two of which were collected by J. Reinhardt in the nineteenth century. *Phyllomys brasiliensis* frequently has been confused with the recently named *P. pattoni* (Emmons et al., 2002), but the description and measurements of *P. brasiliensis* given by Reinhardt (1849) and Winge (1887) are rich in detail and provide unambiguous distinctions between these two species: "aristiforms grayish at the base, orange distally and grayish-brown at the tip, ending in a very fine, curved hairlike point" (Reinhardt, 1849, p. 113, translation from the original Danish); "tail hairs get longer distally, forming a thick tuft at the tip" (Winge, 1887, p. 81, translations from the original Danish). The specimen used in the phylogenetic analysis presented above was collected recently by M. A. Sábato and is the only record in the twentieth century.

Specimens examined (Total = 1): Minas Gerais: Fazenda Santa Cruz (AP 48).

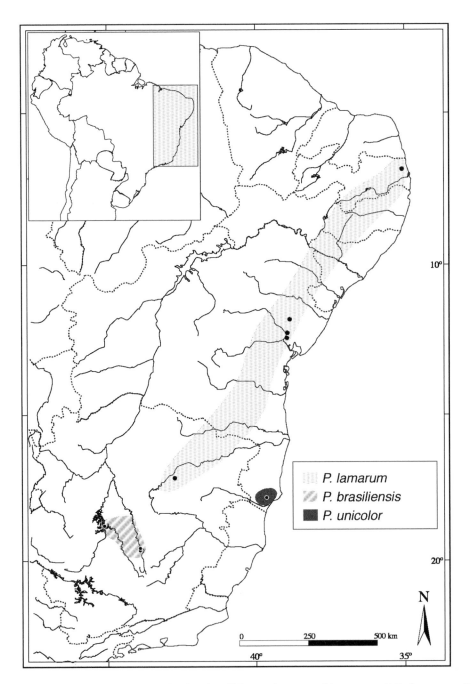

FIGURE 27. Map showing collecting localities and geographic range of *P. lamarum, P. brasiliensis,* and *P. unicolor.*

Phyllomys nigrispinus (Wagner, 1842)

Type material: The holotype is the male NMW B 918, a skin only, collected by Johann Natterer on 7 June 1819 (field number JN 52) at "Ypanema" (Wagner, 1842). The type locality as amended by Emmons et al. (2002) is as follows: Floresta Nacional de Ipanema, 20 km NW Sorocaba, São Paulo, Brazil, 23°26'S 47°37'W, elevation 550 m to 970 m.

Diagnosis: A medium-sized species presenting substantial morphological variation within and among populations. Dorsum spiny, reddish brown streaked with black, clothed with relatively narrow (1 mm), medium-length (27 mm), black aristiform hairs, with thin, whip-like tips (Fig. 5F). Ventral color extremely variable, ranging from buffy-white to yellowish gray, hairs usually white at base. Palatine width equal to or larger than tooth width at M1, toothrows parallel. Supraorbital ridges well developed, interorbital region diverging posteriorly; postorbital process absent or inconspicuous (Fig. 24E). Zygomatic arch slender to moderately robust, maximum height between 1/4 and 1/3 of jugal length. Postorbital process of zygoma rounded or spinose, formed mainly by the jugal. Mastoid process long, extending below the horizontal midline of the external auditory meatus. Mesopterygoid fossa reaches last laminae of M2 or first of M3; wide, forming an angle of about 60 degrees. Incisive foramen ovate. Ventral root of the angular process of mandible does not deflect laterally. Upper incisors slightly opisthodont.

Karyotype: Lena Geise (pers. comm.) found 2n = 52 in specimens from the state of Rio de Janeiro (FS 12-03, 12-30).

Distribution and habitat: *Phyllomys nigrispinus* is found in southeastern Brazil from the states of Rio de Janeiro (ca. 22°S) to Paraná (ca. 26°S), mainly along the coastal zone but extending inland to western São Paulo (ca. 50°W, Fig. 28). Most museum records are in areas of broadleaf evergreen rainforest, but this species also occurs in semideciduous forests.

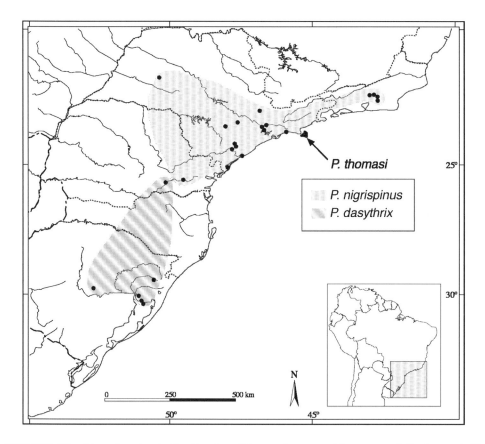

FIGURE 28. Map showing collecting localities and geographic range of *P. thomasi, P. nigrispinus*, and *P. dasythrix*.

Remarks: The level of morphological variation observed in *P. nigrispinus*, especially in skull characters, is higher than in any other species of *Phyllomys*. Even within a single locality, Interlagos, where A. M. Olalla collected 23 specimens, there is significant variation, some skulls being typically short and wide, while others are long and narrow. I have decided to keep all specimens in *P. nigrispinus* since the holotype is a skin only and the number of specimens available at this time precludes any assessment of geographic variation. Future studies, based on a larger number of samples and genetic data, may reveal more than one species in this group. See also remarks under *P. kerri*, below.

Specimens examined (Total = 50): **Paraná**: Guajuvira (MZUSP 6431), **Rio de Janeiro**: Fazenda Alpina, Teresópolis, (MNRJ 31522), FS 12, Cachoeiras de Macacu (FS 12-03, 12-30), FS 6, Cachoeiras de Macacu (FS 6-43), Teresópolis

(MNRJ 6440, 6441, 6442, 6443, 6444, 6445), **São Paulo**: Barra de Icapara (MZUSP 25862, 25863), Ilha do Cardoso, Cananéia (MZUSP 27755), Interlagos (MZUSP 10311, 10312, 10316, 10317, 10318, 10319, 10320, 24939, 24941, 25853, 25854, 25855, 25856, 25857, 25858, 25859, 25860, 25861, 26712, 26713, 26715, 26716, 26721), Itapetininga (MZUSP 175), Itatiba (MZUSP 664, 665, 666), Rio Guaratuba, Santos (UFMG 948), São Paulo (MZUSP 1949, 1950, 1951, 1952, 1953, 1954), Taboão da Serra (MZUSP 26652), Vanuire (MZUSP 3738).

Phyllomys unicolor (Wagner, 1842)

Type material: The holotype is SMF 4319, collected by J. Freyreiss prior to 1824. It consists of a mounted skin and damaged skull removed from the mount in 1998. The type locality was restricted by Emmons et al. (2002) to Colônia Leopoldina (now Helvécia), 50 km SW Caravelas, Bahia, Brazil, 17°48'S 39°39'W, elevation 59 m.

Diagnosis: A very large species with short (less than 2.0 cm on dorsal mid-rump) and stiff narrow hairs of uniform color throughout their length. Color uniform pale rusty red-brown dorsally grading to a buffy venter. Tail rust colored, completely covered with long hairs (ca. 0.5 cm), completely hiding scales and becoming longer towards tip. Feet broad but not long (hindfeet length = 40.5 mm), with stout claws; yellowish above. Ears nearly naked, short (ear length = 16.2 mm), with tuft on anterior rim. Cranium massively built, flat dorsally in lateral profile. Cheekteeth large (maxillary toothrow length = 13.85 mm), toothrows parallel, rostrum short and robust (Fig. 23B). Jugals broad dorsoventrally, jugal fossa deeply concave, with strong, beaded ventral lip; tip of fossa reaches anteriorly into the ventral, maxillary-jugal suture. Tympanic bullae conspicuously inflated; mastoid process extending to the lower edge of external auditory meatus. Mesopterygoid fossa reaches to level of posterior edge of first lamina of M3; angle of fossa is wide, 60 degrees. Incisive foramen ovate. Upper incisors broad, orange, and orthodont. Postorbital process of zygoma rounded, formed mainly by jugal. Ventral root of the angular process of mandible deflected laterally.

Karyotype: Unknown.

Distribution and habitat: *Phyllomys unicolor* is known only from the type locality in the southernmost part of Bahia state, an area of broadleaf evergreen rainforest close to sea level (Fig. 27).

Remarks: This species is known only from the holotype collected before 1824, which I did not examine. My comments here are thus restricted to the examination of photographs and notes taken by D. Kock and L. H. Emmons. Emmons et al. (2002) were the first to accurately redescribe *P. unicolor* since Wagner's original description. Moojen (1952) mentioned *P. unicolor* in his book, but his description does not match that of the holotype. In a previous work (Moojen, 1950, p. 491, translation from the original Portuguese), he compared *P. kerri* with the animal he called *P. unicolor* stating that "we [MNRJ?] have only one male *E. unicolor.*" As pointed out by Emmons et al. (2002), there are very few similarities between the holotypes of *P. unicolor* and *P. kerri*. I was unable to find any specimen at the MNRJ either identified as *P. unicolor* or matching Moojen's (1952) short description, and I assume that his "*P. unicolor*" refers to some other species.

Specimens examined: None.

Phyllomys dasythrix Hensel, 1872

Type material: The skull-only ZMB 38800 is the lectotype, and ZMB 38794, 38795, 38796, 38797, 38798, 38799, 38801, 38802, 38803, 38804, 11651, and BMNH 1.12.3.1 are paralectotypes designated by Emmons et al. (2002). Type specimens were collected by R. Hensel in "Rio Grande do Sul, Süd-Brasiliens" (Hensel, 1872), and the type locality was restricted to Porto Alegre, Rio Grande do Sul, Brazil, 30°04'S 51°07'W, by Emmons et al. (2002).

Diagnosis: A medium-sized species with soft pelage. Aristiforms on rump long (26 mm) and very fine (0.2 mm), paler at the base and blackish distally, with thin whip-like tips (Fig. 5K). Tail moderately covered with brown hairs to the tip, without a tuft. Cheekteeth large, palatine width less than tooth width at M1, and toothrows parallel. Supraorbital ridges weakly developed, interorbital region slightly divergent posteriorly, sometimes with a small postorbital process (Fig. 24A). Zygomatic arch robust, maximum height approximately 1/3 of jugal length. Mastoid process short, extending to the horizontal midline of the external auditory meatus. Mesopterygoid fossa narrow, forms an angle of 45–60 degrees, reaches last lamina of M2. Incisive foramen ovate. Ventral root of the angular process of mandible does not deflect laterally. Upper incisors orthodont.

Karyotype: Alexandre Christoff (pers. comm.) found 2n = 72 in a specimen from Rio Grande do Sul (AC 629).

Distribution and habitat: The known range of *Phyllomys dasythrix* extends from the southern part of the state of Paraná (ca. 26°S) to Rio Grande do Sul (ca. 30°S, Fig. 28). It is found in areas of semideciduous and perhaps Araucaria forests inland, usually below 800 m elevation.

Remarks: After describing *P. medius*, Thomas (1909, p.) compared it with *P. dasythrix*: "The smallest [species], *L. dasythrix*, Hensel, is represented by one of the original typical *skulls* [my italics] from Rio Grande do Sul, and two dealers' specimens from the island of Santa Catherina." It turns out that Thomas was writing about two different species: *Phyllomys dasythrix* and what I call here *P.* aff. *dasythrix* (see account below). I have not examined the BMNH material, but the notes taken in February 1999 by J. L. Patton are clear. On the specimens from Santa Catarina (BMNH 50.7.8.24, 7.1.1.174, 50.7.8.25) he wrote: "fur coarse, aristiform hairs stiff and broad (1.04 mm)"; while on a specimen from Palmira, Paraná (BMNH 0.6.29.20), he wrote: "dorsal fur soft, no aristiform spines... longest hairs 13 mm; not very different from Sta. Catarina specimens, but dorsal fur completely different." My deduction is that Thomas never saw the skin of a true *P. dasythrix*, which is an entirely soft-furred animal. Since the skulls of *P. dasythrix* and *P.* aff. *dasythrix* are almost identical, and he had only one skull of the former, he understandably lumped all specimens he had under *P. dasythrix*. Support for this conjecture comes from his statement years later that "All the species of the genus [*Phyllomys*] are spiny, for it now proves that the non spiny species deserve generic separation" (Thomas, 1916a, p. 296). Support also comes from Ellerman (1940, p. 109) also working at the BMNH, who probably never examined a skin of *P. dasythrix* either, since he stated that in his "*dasythrix* group" (including *P. dasythrix* and *P. lamarum*) "the spines are strong." He was probably referring to the same specimens of *P.* aff. *dasythrix* cited by Thomas. See also remarks under *P.* aff. *dasythrix* below.

Specimens examined (Total = 13): **Rio Grande do Sul**: Bairro Agronomia, Porto Alegre (AC 627, 628), Itapuã, Viamão (MCNFZB 44, 46, 47, 48), Pinheiros, Candelária (MNRJ 6238), Porto Alegre (ZMB 4278, 8237, 38794, 38799, 38800), São Francisco de Paula (MNRJ 21503).

Phyllomys thomasi (Ihering, 1897)

Type material: The skin and skull MZUSP 47, male, was designated as the lectotype, and specimens MZUSP 45, 51, 223, 526, 527, 532; FMNH 41360; BMNH 99.8.12.1, and 2.8.25.2 are paralectotypes (Emmons et al., 2002). They were collected in 1896 at "Ilha de São Sebastião" (Ihering, 1897), type locality amended by Emmons et al. (2002) to Ilha de São Sebastião, São Paulo, Brazil, 23°46'S 45°21' W.

Diagnosis: The largest species in the genus. Dorsal pelage spiny, reddish brown streaked with black, darker on mid-dorsum. Venter cream to light gray, grading gradually from sides. Aristiforms on the rump long (33 mm), relatively narrow (0.7 mm), gray-brown proximally and black distally, with a thin whip-like tip (Fig. 5H). Tail robust, covered with dark brown hair to its tip. Cheekteeth wide, palatine width equal to or larger than tooth width at M1, toothrows parallel to slightly divergent posteriorly. Supraorbital ledges well developed; interorbital region nearly parallel, with inconspicuous postorbital processes (Fig. 24C). Zygomatic arch robust, maximum height more than 1/3 of jugal length. Postorbital process of zygoma usually rounded, formed mainly by jugal. Mastoid process short, extending to the midline of the external auditory meatus. Mesopterygoid fossa wide, forming an angle of about 60 degrees; reaches the posterior lamina of M2. Incisive foramen teardrop shaped. Ventral root of the angular process of mandible does not deflect laterally. Upper incisors slightly opisthodont.

Karyotype: Unknown.

Distribution and habitat: Endemic to the island of São Sebastião, off the coast of São Paulo (336 km^2, Fig. 28). The island has rugged topography, going from sea level to 1379 meters, and the dominant vegetation is broadleaf evergreen rainforest (see also Ihering, 1897; Olmos, 1996, 1997, for habitat description).

Remarks: Ihering based his original description of *P. thomasi* on the fact that the only specimen he had examined had no tail (Ihering, 1897). After he had inspected the remaining specimens collected in the original series, some with tail and some without, he concluded that it was conspecific with *P. nigrispinus* (Ihering, 1898). *Phyllomys thomasi* is however distinct from *P. nigrispinus* by its much larger sizebut is otherwise very similar in general morphology. It is also very similar to *P. kerri* in cranial shape but is proportionally larger in all skull measurements (see discussion in Cranial Morphometrics, below).

Specimens examined (Total = 19): **São Paulo**: Ilha de São Sebastião (MNRJ 6448, MZUSP 45, 47, 51, 223, 526, 527, 532, 2148, 2149, 2151, 3197, 3199, 6433, 25816), Ilhabela, Ilha de São Sebastião (MZUSP 3198), Trilha da Água Branca, Ilha de São Sebastião (MZUSP 29017), specific locality unknown [probably Ilha de São Sebastião] (MZUSP 535, USNM 296336).

Phyllomys medius (Thomas, 1909)

Type material: The holotype is the skin and skull BMNH 3.4.1.84, an adult female, collected on 18 September 1901 by A. Robert (field number 864) at "Roça Nova, Serro [*sic*] do Mar, Parana, S[outh]. Brazil. Alt. 1000 m" (Thomas, 1909). Type locality amended by Emmons et al. (2002) to Roça Nova, Serra do Mar, Paraná, Brazil, 25°28'S 49°01'W, elevation 1000 m.

Diagnosis: A medium-sized *Phyllomys* with stiff but relatively soft pelage. Dorsum dark brown sprinkled with black, darkest on mid-dorsum. Aristiforms on rump very long (36 mm), thin (0.4 mm), black distally, with a thin whip-like tip (Fig. 5I). Ventral hair distinctly gray-based with fulvus tips. Skull robust and long; auditory tympanic bullae small; incisive foramen teardrop shaped and small. Teeth narrow, palatine width equal to tooth width at M1; toothrows parallel. Supraorbital ledges well developed, interorbital region parallel sided or divergent posteriorly, postorbital process absent or inconspicuous (Fig. 24G). Zygomatic arch robust, maximum height more than 1/3 of jugal length. Postorbital process of zygoma spinose, formed mainly by jugal. Mastoid process short, extending to the horizontal midline of the external auditory meatus. Mesopterygoid fossa reaches the last lamina of M2 and narrow, forming an angle of about 45 degrees. Ventral root of the angular process of mandible does not deflect laterally. Upper incisors opisthodont.

Phallus: Glans penis of one adult specimen examined (EDR 8) elongated, measuring 12 mm in length and 3 mm in maximum width (Fig. 6C). Terminal portion is cone shaped, ending in a pointed bacullar papilla. Urethral lappet is evident ventrally as a flat inverted lyre-shaped surface, bordered by small lateral skin folds.

Karyotype: Sbalqueiro et al. (1988) reported 2n = 96 for two specimens from Santa Catarina, listed as "*Echimys dasythrix.*" I personally examined one of these specimens (MS 92, I. Sbalqueiro, pers. comm.), and it is *P. medius.*

Distribution and habitat: *Phyllomys medius* occurs from the states of Minas Gerais and Rio de Janeiro in the southeast (ca. 21°S) to Rio Grande do Sul in the south (ca. 30°S, Fig. 29). It is found mainly along the coast in areas of broadleaf evergreen rainforest, but it extends to the west into Araucaria forests in the state of Paraná (ca. 52°W). The two northern records (Terezópolis, RJ, and Poços de Caldas, MG) are from above 1000 m elevation, while the southern records are from lower elevations, suggesting preference for cooler climates (see also Voss, 1993, who observed the same trend in the sigmodontine rodent genus *Delomys*). Davis

(1945) provides a detailed description of the habitat of *P. medius* at Terezópolis, RJ.

Specimens examined (Total = 26): **Minas Gerais**: Alto da Consulta, Poços de Caldas (MNRJ 31561), **Paraná**: Porto Camargo, Rio Paraná (MZUSP 7716), Represa de Foz do Areia (MZPUCPR 129, EDR 8, 11, FK 21, 22, 23, 24, 25), Rio Paracaí (MZUSP 7717), Salto Morato, Guaraqueçaba (MHNCI 3033), **Rio de Janeiro**: Fazenda Comari, Teresópolis (MNRJ 6237, 6240, 6742), **Rio Grande do Sul**: Banhado do Pontal, Triunfo (MCNFZB 391), **Santa Catarina**: Florianópolis (MNRJ 31568, 31569, 31570, 31571, 31572), Lagoa da Conceição, Florianópolis (MS 92), Praia dos Açores, Ilha de Santa Catarina (MNRJ 48864), Sítio Garça Branca, Florianópolis (AX 30), **São Paulo**: Barra do Ribeirão Onça Parda (MZUSP 10629), **unknown locality** (MZUSP 528).

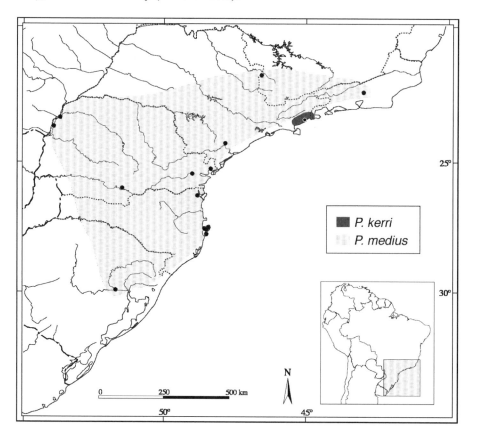

FIGURE 29. Map showing collecting localities and geographic range of *P. kerri* and *P. medius*.

Phyllomys lamarum (Thomas, 1916)

Type material: The holotype is the skin and skull BMNH 3.9.5.96, an adult female collected by A. Robert on 4 March 1903 (field number 1414) at "Lamaraõ [*sic*], Bahia, about 70 miles N.W. of Bahia city [now Salvador]. Alt. 300 mm [*sic*]." (Thomas, 1916a). Type locality amended by Emmons et al. (2002) to Lamarão, about 70 miles NW Salvador, Bahia, Brazil, 11°47'S 38°53'W, elevation 300 m. In his original description Thomas (1916a) mentioned "thirteen specimens," but there are 12 others in London besides the holotype (BMNH 3.9.5.1, 3.9.5.92, 3.9.5.93, 3.9.5.94, 3.9.5.95, 3.9.5.97, 3.9.5.98, 3.9.5.99, 3.9.5.101, 3.9.5.102, 3.9.5.103, 3.9.5.104), one in Chicago (FMNH 35356), one in Paris (MNHN 754, 1821A), and two in Copenhagen (UZMC 1281, 1282), and it is not clear which specimens Thomas had in hand when he wrote the description.

Diagnosis: A small- to medium-sized species with conspicuous spines. Dorsum is yellow-brown with a speckled pattern given by the short (24 mm), broad (1.3 mm) aristiforms (Fig. 5B). Spines pale at the base, darkening towards the tip where they become orange, ending in dark, thin, whip-like tips. Venter pale brown with white patches, to pure white, with fulvus lateral line. Tail relatively slender, slightly shorter than or equal to head and body, thinly covered with pale brown hairs, scales visible to the eye along entire length. Cheekteeth wide, palatine width equal to or shorter than tooth width at M1, toothrows slightly divergent at either end. Supraorbital ledges well developed. Interorbital region almost straight, diverging posteriorly, postorbital process absent or inconspicuous (Fig. 23D). Zygomatic arch relatively robust, maximum height less than or equal to 1/3 of jugal length. Postorbital process of zygoma spinose, usually with no contribution from squamosal. Mastoid process short, extending to the midline of the external auditory meatus. Mesopterygoid fossa reaches last lamina of M2; wide, forming an angle of more than 60 degrees. Incisive foramen ovate. Ventral root of the angular process deflected laterally. Ventral spine present near the posterior junction of the mandibular rami. Upper incisors orthodont.

Distribution and habitat: The range of *Phyllomys lamarum* extends from the state of Paraíba in the northeast (ca. 7°S) to northern Minas Gerais in the southeast (ca. 17°S, Fig. 27). It is found mainly inland in semideciduous forests but reaches the coastal area in Paraíba. The three clusters of records for *P. lamarum* are about 700 km apart, and those populations may actually be disjunct.

Karyotype: Unknown.

Remarks: As mentioned above under *P. blainvilii*, Moojen (1952) had the two species reversed. Specimens of *P. lamarum* from the arid parts of the range, such as the type locality Lamarão, are paler than specimens from moister areas, for example, in Minas Gerais.

Specimens examined (Total = 31): **Bahia**: 9 km SE Feira de Santana (MNRJ 11260), Fazenda Boa Vista, Feira de Santana (MNRJ 21659), Fazenda Estiva, Feira de Santana (MNRJ 21650, 21655, 21656, 21657, 21658), Fazenda Estrada Nova, Feira de Santana (MNRJ 21660), Fazenda Morro, Feira de Santana (MNRJ 21654), Fazenda Quituba, Feira de Santana (MNRJ 21662, 21663, 21664, 21665), Fazenda Salgado, Feira de Santana (MNRJ 21661), Fazenda Terra Nova, Feira de Santana (MNRJ 21666, 21667, 21668, 21669), Fazenda Três Riachos, Feira de Santana (MNRJ 21652, 21653), Lamarão, about 70 miles [112 km] NW Salvador (FMNH 35356), São Gonçalo, 30 km SW Feira de Santana (MNRJ 11259), specific locality unknown [probably Feira de Santana] (MNRJ 2F.691), **Minas Gerais**: Estação Ecológica de Acauã (LC 73), **Paraíba**: Camaratuba, Mamanguape (MZUSP 8413, 8415, 8416, 8417, 8418), Uruba, Mamanguape (MZUSP 8414), **unknown locality** (MNRJ 21553).

Phyllomys kerri (Moojen, 1950)

Type material: The holotype is the skin and skull MNRJ 6241, an adult female collected on 13 June 1941 by G. Dutra at "Ubatuba, S[ão]. Paulo, Brasil" (Moojen, 1950). Type locality restricted by Emmons et al. (2002) to Estação Experimental de Ubatuba, Ubatuba, São Paulo, Brazil, 23°25'S 45°07'W.

Diagnosis: A medium-sized species. Dorsum spiny, reddish brown streaked with black. Aristiform hairs on rump black, 27 mm in length and 1 mm in width, with thin, whip-like tips (Fig. 5G). Ventral color yellowish orange, hairs usually gray at the base. Skull long and narrow, palatine width equal to or larger than tooth width at M1, toothrows nearly parallel. Supraorbital ridges well developed, interorbital region diverging posteriorly, with an inconspicuous postorbital process (Fig. 24D). Zygomatic arch robust, maximum height approximately 1/3 of jugal length. Postorbital process of zygoma spinose and formed mainly by jugal. Mastoid process short, reaching the horizontal midline of the external auditory meatus. Mesopterygoid fossa reaches last laminae in M2; wide, forming an angle of nearly 60 degrees. Incisive foramen ovate. Ventral root of the angular process of mandible does not deflect laterally. Upper incisors slightly opisthodont.

Karyotype: Unknown.

Distribution and habitat: Known only from Ubatuba, on the northern coast of São Paulo state (Fig. 29). The vegetation in the area is broadleaf evergreen rainforest and the elevation is close to sea level.

Remarks: Although the original series collected by G. Dutra between 13 June and 27 July 1941 consists of three specimens, Moojen (1950) mentioned only the holotype in the description. As pointed out above under *P. nigrispinus*, the morphological differences between *P. kerri* and *P. nigrispinus* are likely to be within the range of variation found within specimens currently recognized as *P. nigrispinus* (see also Cranial Morphometrics, below). Nevertheless, I was able to find diagnostic characters of *P. kerri*, such as short maxillary toothrow, broad rostrum, wide interorbital region, and reduced bulla. Consequently, I have decided to provisionally maintain *P. kerri* as a species, at least until enough specimens of both taxa become available to allow adequate comparisons.

Specimens examined (Total = 3): **São Paulo**: Estação Experimental, Ubatuba (MNRJ 6241), Ubatuba (MNRJ 5463, 5464).

Phyllomys pattoni Emmons et al., 2002

Type material: The holotype is an adult female MNRJ 62391, collected by Yuri Leite (field number 197) on 14 April 1998, consisting of skin, skull, carcass and liver sample (Emmons et al., 2002). The type locality is Mangue do Caritoti, Caravelas, Bahia, Brazil, 17°43'30"S 39°15'35"W, sea level. There are six paratypes, which have been deposited in MZUSP (YL 198, 199, 200, 201), MVZ (197621), and USNM (YL 196).

Diagnosis: A medium-sized species and the most heavily spined tree rat in the genus. Aristiforms on rump average 23 mm in length and 1.0 in width. Dorsal spines light gray at the base, gradually darkening towards black medially, terminating in a blunt orange tip (Fig. 5D). Most aristiforms on rump lack the characteristic whip-like tip present in all other species of *Phyllomys* (Emmons et al., 2002). Dorsum dark brown, with a speckled aspect given by the rusty-tipped spines. Tail slightly shorter to slightly longer than head and body length and covered with fine hair to the tip, not hiding the scales. Skull robust and relatively broad (Fig. 23C). Maxillary toothrows relatively short and narrow, palatine width equal to tooth width at M1. Supraorbital ledges well developed, beaded. Interorbital region divergent posteriorly without postorbital process. Zygomatic arch robust, maximum height approximately 1/3 of jugal length. Postorbital process of zygoma spinose, formed by jugal and squamosal. Mastoid process long, extending below the horizontal midline of the external auditory meatus. Incisive foramen bullet shaped. Ventral root of the angular process of mandible deflected laterally. Upper incisors orthodont.

Phallus: The glans of two adult specimens examined (YL 198, 199) is bullet shaped and robust (Fig. 6D). The largest (YL 198) measures 11 mm in length and 4 mm in maximum width. A 1.5-mm-long bacular papilla is present at distal end. Urethral lappet is present and notably wrinkled. Ventral margin of the intromittent sac V-shaped and delimited by two large lateral folds. Intromittent sac is deep (ca. 4 mm) without spikes on its floor. Baculum of one adult specimen examined (YL 199) occupies 2/3 of the glans. It is rod shaped and long, the bone measuring 6.5 mm and connecting distally to a 2.8-mm-long teardrop-shaped cartilaginous tip (Fig. 6E). The bone is wider than deep, convex dorsally and concave ventrally. The maximum width at the sub-proximal enlargement is 1.52 mm, and the minimum width at the sub-terminal constriction is 0.67 mm.

Distribution and habitat: *Phyllomys pattoni* ranges from the state of Paraíba (ca. 7°S) in the northeast to São Paulo (ca. 23°S) in the southeast, mainly along the coast (Fig. 26). It is found chiefly in broadleaf evergreen rainforests and associated habitats, such as mangroves, from sea level to 1000 m.

Karyotype: Zanchin (1988) reported 2n = 80 for a specimen of *P. pattoni* (AL 2398 = UFPB 346) identified as "*Echimys* sp.," collected in the state of Espírito Santo, and Severo (1998) recorded 2n = 80, FN = 112 for the same specimen, identifying it as "*Echimys thomasi*." According to L. Geise (pers. comm.), specimens from the state of Rio de Janeiro (MNRJ 42978, FS 11-52) have 2n = 72, FN = 114.

Remarks: As noted above, this species was incorrectly associated with Lund's *P. brasiliensis* for more than a century (e.g., Burmeister, 1854; Moojen, 1952; Thomas, 1916a).

Specimens examined (Total = 56): **Bahia**: Aritaguá-Urucutuca, Ilhéus (MNRJ 21517), Fazenda Almada, Ilhéus (MNRJ 11253), Fazenda Monte Castelo, Ilha da Cassumba (YL 198, 199, 200, 201), Fazenda Pirataquicê, Ilhéus (MNRJ 10452, 10453, 11251, 11256, 11257, 11258), Helvécia, Nova Viçosa (HGB 36), Itabuna (MNRJ 33515), Mangue do Caritoti, Caravelas (MNRJ 62391, YL 195, 196), Mata Fortuna, Itabuna (MNRJ 11254), São Felipe (MNRJ 22264, 22265), **Espírito Santo**: Estação Biológica de Santa Lúcia (MBML 2011, 2032, 2047), Fazenda Santa Terezinha, 33 km NE Linhares (LC 32), Hotel Fazenda Monte Verde, 24 km SE Venda Nova do Imigrante (UFPB 346), Mata da Torre, Vitória (MBML 162), Parque Estadual da Fonte Grande (MBML 316), Povoação, Linhares (MBML 1856), Reserva Biológica de Duas Bocas (MBML 226), Rio São José (MNRJ

8276), **Minas Gerais**: Fazenda Montes Claros, Caratinga (GABF 97P), Fazenda Paraíso, Além Paraíba (MNRJ 10454), Fazenda São Geraldo, Além Paraíba (MNRJ 4077), Mata Paraíso, Viçosa (UFV 385, 696), Silvicultura, Viçosa (UFV 379), **Paraíba**: João Pessoa (UFPB 774), **Pernambuco**: Dois Irmãos, Recife (MNRJ 8195), **Rio de Janeiro**: Fazenda Alpina, Teresópolis (MNRJ 1933), Fazenda Rosimery, Município de Cachoeiras de Macacu (FS 11-52), Fazenda São José da Serra, Bonsucesso (MVZ 183139), Fazenda União, Casimiro de Abreu (MNRJ 42978), Ilha Grande (MNRJ 31562, 31566, MZUSP 26718), Nova Friburgo (MNRJ 31564, 31567), Saco de São Francisco, Niterói (MNRJ 6449), Santa Cruz, estrada Rio-Petrópolis (MNRJ 21508), Teresópolis (MNRJ 2232, 2237, 2239, 2240), Tijuca, Rio de Janeiro (MNRJ 11252), Vila Dois Rios, Ilha Grande (IG 34), **São Paulo**: Piquete (MZUSP 138).

Phyllomys lundi Leite, present report

Type material: See species description above.

Diagnosis: See species description above and figures 5E and 23A.

Karyotype: Unknown.

Distribution and habitat: Known only from two localities 200 km apart in southern Minas Gerais and Rio de Janeiro, both in broadleaf evergreen rainforest habitat (Fig. 30). See species description above for details.

Specimens examined (Total = 2): **Minas Gerais**: Fazenda do Bené, 4 km SE Passa Vinte (MNRJ 62392), **Rio de Janeiro**: Reserva Biológica de Poço das Antas (RBPDA 2228).

Phyllomys mantiqueirensis Leite, present report

Type material: See species description above.

Diagnosis: See species description above and figures 5L and 24F.

Karyotype: Unknown.

Distribution and habitat: Known only from the type locality in the Serra da Mantiqueira, Minas Gerais, an area of mixed montane rainforest at 1800 m of altitude (Fig. 30). See species description above for details.

Specimens examined (Total = 1): **Minas Gerais**: Fazenda da Onça, 13 km SW Delfim Moreira (MNRJ 62393).

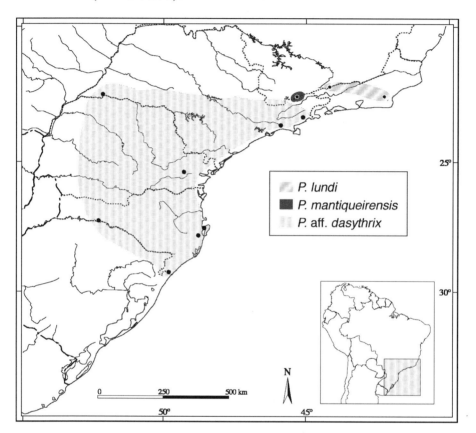

FIGURE 30. Map showing collecting localities and geographic range of *P. lundi, P. mantiqueirensis*, and *P.* aff. *dasythrix*.

Phyllomys aff. *dasythrix*

Comment: This taxon is unnamed and will be described elsewhere (A. Christoff and Y. Leite, in prep.).

Diagnosis: Similar to *P. dasythrix* in size and external appearance, but with coarser fur. Aristiforms on rump medium in length (26 mm) and more than 0.6 mm in width, wider and paler at the base, thining gradually towards the black, thin whip-like tip (Fig. 5J). Tail moderately covered with brown hairs; without a tuft. Skull very similar to *P. dasythrix*: cheekteeth large, palatine width usually less than tooth width at M1, and toothrows parallel. Supraorbital ridges weakly developed, interorbital region slightly divergent posteriorly, sometimes with a small postorbital process (Fig. 24B). Zygomatic arch robust, maximum height approximately 1/3 of jugal length. Postorbital process of zygoma spinose, formed mainly by jugal. Mastoid process short, extending to the horizontal midline of the external auditory meatus. Mesopterygoid fossa is narrow, forms an angle of 45–60 degrees, and reaches last lamina of M2. Incisive foramen ovate. Ventral root of the angular process of mandible does not deflect laterally. Upper incisors orthodont.

Distribution and habitat: *Phyllomys* aff. *dasythrix* ranges from the state of Rio Grande do Sul (ca. 29°S) to São Paulo (ca. 23°S) and perhaps Minas Gerais (Fig. 30). Specimens have been recorded in broadleaf evergreen forests on the coast, as well as semideciduous forests well inland (ca. 52°W). It appears to be found usually above 800 m elevation.

Karyotype: Alexandre Christoff (pers. comm.) found 2n = 92 in specimens from Rio Grande do Sul. Yonenaga (1975) reported 2n = 90 for a female "*Echimys* sp." from Casa Grande, São Paulo, collected by members of the Instituto Adolfo Lutz and made available by Oscar Souza Lopes (see Yonenaga, 1975, p. 270), but without making any reference to a voucher specimen. She also mentioned that "the identification of the specimens was performed by Dr. Ronald H. Pine, from the Smithsonian Institution, Washington, D.C." During my visit to the Smithsonian, I examined a male *Phyllomys* (USNM 460569) from Estação Biológica de Boracéia, São Paulo, collected by Oscar Souza Lopes in 1966, belonging to the species I call *P.* aff. *dasythrix* here. Considering that members of the Instituto A. Lutz collected specimens between 1962 and 1972 at both Casa Grande and Boracéia, which are less than 10 km apart (see also Voss, 1993), I conjecture that the specimen karyotyped by Yonenaga belongs to the same series of USNM 460569 and probably the same species, but, without reference to a voucher, it is impossible to make a positive association.

Remarks: As mentioned earlier under *P. dasythrix*, Thomas (1909) failed to recognize the differences between *P. dasythrix* and *P.* aff. *dasythrix* probably because he only had a skull of the former available and the most obvious distinction among them is the quality of the fur, which is soft in the former and spiny in the latter. James Patton (pers. comm.) noted this difference when he examined specimens in the BMNH. Olmos (1997, p. 134) also noticed the singularity of *P.* aff. *dasythrix*: "two *Nelomys* at the MZUSP from the deciduous forests of western São Paulo may well represent an undescribed species."

Specimens examined (Total = 9): **Paraná**: Parque Barigüi, Curitiba (MHNCI 2599), **Rio Grande do Sul**: Usina Hidrelétrica de Itá (AC 639, 640), **Santa Catarina**: Serra do Tabuleiro (JCV 28), **São Paulo**: Estação Biológica de Boracéia (USNM 460569), Parque Estadual da Serra do Mar, Núcleo Santa Virgínia (NSV 160599), Teodoro Sampaio (MZUSP 25819, 8885), **unknown locality** [probably from the state of Minas Gerais] (YL 194).

Unidentified Specimens

I was unable to identify eight specimens of *Phyllomys* represented by skins only or damaged skulls, or very young individuals with no locality data: **Minas Gerais**: Bacia do Rio das Velhas, Baldim (UFMG 867), Serra do Ibituruna, Governador Valadares (UFMG 951), **unknown locality** (MNRJ 1930, 31521, 31527, 31529, USNM 484508, 485175).

Fossil Record

The only fossils of *Phyllomys* known are from the Late Pleistocene deposits in the limestone caves of Lagoa Santa, collected and studied by P. W. Lund (1839a, b; 1840a, b, c) and housed at the paleontological collection of the UZMC in Copenhagen. Lund first mentioned two species of *Phyllomys*, the extant *P. brasiliensis* and the extinct *P.* aff. *brasiliensis*, without describing them (Lund, 1839a, 1840a), but illustrating their upper cheekteeth (Lund, 1840a, plate 21, figs. 12 and 13). In his next accounts (Lund, 1839b, 1840b), he listed the same two *Phyllomys* and added a third taxon: *Lonchophorus fossilis*. He did not describe *L. fossilis* but provided an illustration of the lower PM and M1 next to a drawing of lower PM and M1 of *P.* aff. *brasiliensis* (Lund, 1840b, plate 25, figs. 9 and 10).

Louise H. Emmons examined Lund's material and concluded that there are indeed three species of Atlantic tree rats among the specimens from which Lund's plate 25, figs. 9 and 10, were drawn, differing mainly by size (see figure 1 in Emmons et al., 2002). However, their identity and relationship with Lund's plates and names are ambiguous (Emmons et al., 2002). The mid sized form seems to be *L. fossilis* and the other two cannot be unambiguously associated with any extant form. Emmons et al. (2002) pointed out that *Lonchophorus* is not distinguishable from *Phyllomys*, and therefore *Lonchophorus* is a junior synonym of *Phyllomys*, and the specimen illustrated may actually represent a *P. brasiliensis*. Consequently, *P. fossilis* is a nomem dubium to be applied only to the holotype (Emmons et al., 2002). The largest toothrow approaches the size found in the largest species (*P. unicolor*, *P. medius*, *P. thomasi*), while the smallest has a toothrow shorter than is found in the smallest extant species (*P. lamarum*, *P. lundi*). Both forms probably represent undescribed, extinct species.

CRANIAL MORPHOMETRICS

Morphometric analyses in *Phyllomys* are severely hampered by the very small number of museum specimens available and the uneven distribution of samples among species. I measured a total of 341 specimens, including individuals of both sexes and all age categories. Most species are represented by 15 specimens or fewer, three species by only one individual (Fig. 31). In addition, even if all specimens are pooled, most localities are represented by single individuals, and only ten localities are the source of more than ten museum specimens. Furthermore, these represent all ages and both sexes, making it impossible to understand ontogenetic and sexual variation, greatly important in other echimyids (e.g., Patton and Rogers, 1983). Evaluation of geographic variation within species is also unfeasible. Trends observed here should be used to direct future studies that include more specimens when they become available.

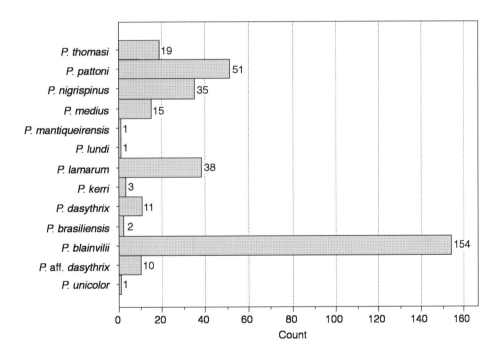

FIGURE 31. Histograms showing the number of specimens used in the morphometric analysis for each species of *Phyllomys*.

Non-geographic Intraspecific Variation in *P. blainvilii*

To assess non-geographic variation, I pooled specimens of *P. blainvilii* from the vicinity of Crato (localities 1–18 in the Gazetteer), because it comprises the largest sample size from a single area: 100 individuals (65 females, 34 males, 1 unknown sex). The ten toothwear age classes were arbitrarily recoded into 5 categories to increase the sample size as follows: very young (age classes 1 and 2; not represented in the Crato samples), young (age classes 3 and 4; 4 individuals), sub-adult (age classes 5 and 6; 11 specimens), adult (age classes 7 and 8; 45 individuals), and old adult (age classes 9 and 10; 24 specimens).

The results of the two-way analysis of variance show no sexual dimorphism in this population of *P. blainvilii* for any of the measurements taken (Table 4). Likewise the interaction between sex and age had virtually no effect on the variation observed. On the other hand, age variation was statistically significant for all measurements, except IFL, MaxB, and CD (Table 5). These results are concordant with those reported for other echimyids, such as *Proechimys brevicauda* (Patton and Rogers, 1983), *Trinomys albispinus* (Pessoa and Reis, 1991), *Proechimys simonsi* and *Mesomys hispidus* (Patton et al., 2000), which showed significant morphometric within-locality variation due to age and none due to sex or the interaction of the two variables. Even when only adults and old adults (toothwear age classes 7–10) were considered, most variables still exhibited statistically significant age variation (Table 5). This confirms the general observation of indeterminant growth recorded in echimyids, as size increases even among adult toothwear age-class animals (Lara et al., 1992; Patton and Rogers, 1983; Patton et al., 2000). Only one measurement (PLb) showed significant sexual dimorphism when just adults and old adults were considered (Table 4), and the combined effect of sex and age was again insignificant.

TABLE 4. Cranial measurements of *Phyllomys blainvilii* from Crato, Ceará. Mean, standard deviation, and sample size for males and females for all ages combined (3–10) and ages 7–10 only. Probability level (p) for sex differences given as: ns (non-significant) = p > 0.05; * = p < 0.05.

	ages 3–10				ages 7–10			
	females	males	p		females	males	p
GSL	48.12 ± 3.01 (58)	47.90 ± 2.54 (33)	ns		49.18 ± 1.68 (37)	48.21 ± 1.98 (26)	ns
NL	13.66 ± 1.35 (62)	13.74 ± 1.29 (32)	ns		13.98 ± 0.94 (39)	14.04 ± 1.14 (24)	ns
RL	16.80 ± 1.53 (60)	16.69 ± 1.32 (32)	ns		17.26 ± 0.92 (38)	16.90 ± 0.96 (24)	ns
OL	13.87 ± 0.81 (65)	13.68 ± 0.65 (34)	ns		14.16 ± 0.45 (42)	13.82 ± 0.50 (26)	ns
RB	6.88 ± 0.42 (63)	6.90 ± 0.39 (34)	ns		7.03 ± 0.36 (40)	7.00 ± 0.37 (26)	ns
IOC	9.80 ± 0.58 (65)	10.06 ± 0.62 (34)	ns		9.96 ± 0.50 (42)	10.13 ± 0.57 (26)	ns
MB	18.77 ± 0.88 (63)	18.75 ± 0.82 (33)	ns		19.02 ± 0.70 (40)	18.94 ± 0.61 (26)	ns
ZB	23.13 ± 1.18 (55)	22.61 ± 1.24 (28)	ns		23.43 ± 0.93 (37)	22.99 ± 0.91 (21)	ns
CIL	43.15 ± 3.32 (59)	42.88 ± 2.32 (33)	ns		44.56 ± 1.76 (37)	43.26 ± 1.72 (26)	ns
BaL	36.94 ± 3.02 (59)	36.60 ± 2.16 (30)	ns		38.20 ± 1.51 (36)	36.93 ± 1.49 (23)	ns
D	9.60 ± 1.00 (64)	9.58 ± 0.85 (34)	ns		9.90 ± 0.72 (41)	9.73 ± 0.73 (26)	ns
MTRL	11.28 ± 0.94 (65)	11.46 ± 0.73 (34)	ns		11.59 ± 0.44 (42)	11.65 ± 0.33 (26)	ns
PLb	8.61 ± 0.81 (65)	8.54 ± 0.73 (34)	ns		8.92 ± 0.49 (42)	8.67 ± 0.52 (26)	*
IFL	3.81 ± 0.38 (53)	3.74 ± 0.42 (31)	ns		3.82 ± 0.36 (32)	3.78 ± 0.42 (24)	ns
BuL	10.11 ± 0.58 (64)	9.90 ± 0.52 (34)	ns		10.31 ± 0.46 (41)	9.94 ± 0.53 (26)	ns
Pla	18.06 ± 1.73 (64)	18.05 ± 1.42 (34)	ns		18.71 ± 1.01 (41)	18.38 ± 1.00 (26)	ns
PPL	19.65 ± 1.06 (59)	19.48 ± 1.00 (32)	ns		20.00 ± 0.87 (38)	19.64 ± 0.90 (25)	ns
MPF	3.19 ± 0.26 (65)	3.18 ± 0.24 (33)	ns		3.25 ± 0.21 (42)	3.23 ± 0.23 (25)	ns
MaxB	7.37 ± 0.36 (63)	7.43 ± 0.32 (34)	ns		7.34 ± 0.33 (41)	7.49 ± 0.33 (26)	ns
OccW	8.33 ± 0.30 (60)	8.37 ± 0.27 (31)	ns		8.38 ± 0.28 (38)	8.37 ± 0.24 (24)	ns
RD	10.30 ± 0.70 (59)	10.18 ± 0.67 (33)	ns		10.51 ± 0.42 (37)	10.29 ± 0.59 (25)	ns
CDM1	13.77 ± 0.74 (64)	13.67 ± 0.78 (34)	ns		13.99 ± 0.48 (42)	13.84 ± 0.54 (26)	ns
CD	18.28 ± 0.73 (43)	18.47 ± 1.18 (22)	ns		18.40 ± 0.58 (24)	18.39 ± 0.96 (15)	ns

TABLE 5. Cranial measurements of *Phyllomys blainvilii* from Crato, Ceará. Mean, standard deviation, and sample size for four age categories. Probability level (p) for age differences given as: ns (non-significant) = p > 0.05; * = p < 0.05; ** = p < 0.01; *** = p < 0.001.

	young (ages 3–4)		sub-adult (ages 5–6)		adult (ages 7–8)		old adult (ages 9–10)		p
GSL	38.55 ± 2.12	(3)	44.69 ± 1.72	(10)	48.04 ± 1.68	(43)	50.22 ± 1.24	(21)	***
NL	10.18 ± 0.41	(4)	12.35 ± 0.70	(11)	13.67 ± 0.89	(44)	14.71 ± 0.88	(20)	***
RL	12.50 ± 0.67	(4)	15.02 ± 0.75	(10)	16.71 ± 0.79	(43)	17.93 ± 0.73	(20)	***
OL	11.36 ± 0.25	(4)	13.02 ± 0.50	(11)	13.85 ± 0.46	(45)	14.36 ± 0.36	(24)	***
RB	6.10 ± 0.30	(4)	6.55 ± 0.26	(11)	6.90 ± 0.32	(44)	7.25 ± 0.32	(23)	***
IOC	8.72 ± 0.42	(4)	9.57 ± 0.64	(11)	9.83 ± 0.46	(45)	10.37 ± 0.46	(24)	***
MB	16.35 ± 0.47	(4)	18.29 ± 0.70	(10)	18.75 ± 0.56	(43)	19.38 ± 0.66	(24)	***
ZB	19.63 ± 0.66	(3)	21.58 ± 1.04	(8)	22.90 ± 0.93	(39)	23.88 ± 0.72	(20)	***
CIL	33.45 ± 1.71	(4)	39.72 ± 1.40	(10)	43.25 ± 1.63	(43)	45.53 ± 1.20	(21)	***
BaL	27.92 ± 1.57	(4)	33.73 ± 1.38	(10)	37.09 ± 1.46	(43)	39.16 ± 0.81	(17)	***
D	6.99 ± 0.62	(4)	8.56 ± 0.47	(11)	9.54 ± 0.64	(45)	10.35 ± 0.62	(23)	***
MTRL	8.64 ± 1.18	(4)	10.33 ± 1.12	(11)	11.63 ± 0.40	(45)	11.59 ± 0.41	(24)	***
PLb	6.24 ± 0.34	(4)	7.73 ± 0.55	(11)	8.62 ± 0.40	(45)	9.19 ± 0.48	(24)	***
IFL	3.33 ± 0.35	(4)	3.60 ± 0.40	(10)	3.79 ± 0.35	(39)	3.81 ± 0.47	(18)	ns
BuL	9.04 ± 0.13	(4)	9.48 ± 0.42	(11)	9.96 ± 0.46	(44)	10.52 ± 0.40	(24)	**
Pla	13.11 ± 0.64	(4)	16.01 ± 0.86	(11)	18.10 ± 0.89	(45)	19.47 ± 0.52	(23)	***
PPL	16.85 ± 0.56	(3)	18.62 ± 0.70	(10)	19.58 ± 0.74	(41)	20.36 ± 0.92	(23)	***
MPF	2.58 ± 0.19	(4)	3.02 ± 0.17	(11)	3.16 ± 0.18	(43)	3.39 ± 0.19	(24)	***
MaxB	7.00 ± 0.22	(4)	7.45 ± 0.35	(11)	7.44 ± 0.32	(44)	7.32 ± 0.36	(23)	ns
OccW	7.83 ± 0.21	(4)	8.28 ± 0.28	(10)	8.33 ± 0.23	(43)	8.50 ± 0.28	(20)	*
RD	8.34 ± 0.57	(4)	9.76 ± 0.62	(10)	10.25 ± 0.45	(43)	10.76 ± 0.43	(20)	***
CDM1	11.42 ± 0.20	(3)	12.77 ± 0.54	(11)	13.69 ± 0.40	(45)	14.38 ± 0.35	(24)	***
CD	17.89 ± 0.77	(4)	18.55 ± 1.33	(10)	18.39 ± 0.76	(28)	18.43 ± 0.69	(12)	ns

Interspecific Variation

I extrapolated the pattern of non-geographic variation in *P. blainvilii* to the remaining species. Thus, I pooled males and females and restricted the following multivariate analyses to individuals of toothwear age category 8 only. Table 6 presents the descriptive statistics for each species.

A total of 119 cases had to be excluded in multivariate analyses when all 23 measurements were included due to missing values. To overcome this limitation, I excluded variables with 18 or more missing cases. In the first principal components analysis including all taxa, I also excluded six other measurements that were missing for the only specimen of *P. unicolor* available. The matrix thus included 9 variables and 146 cases. Table 7 shows the eigenvalues and percent contribution of each component. Nearly all variables load positively and very high on the first axis (PC-1), which accounts for 69.8% of the variation. This first principal component is usually associated with overall size (Bookstein et al., 1985), and it clearly separates the two largest species, *P. thomasi* and *P. unicolor*, from the remainder (Fig. 32). PC-2 accounts for 11.0% of the total, but there are no obvious clusters of species, including when PC-2 was plotted against PC-3, 4, or 5.

TABLE 6. Selected cranial measurements (in millimeters) of adult (age category 8) males and females of *Phyllomys*, including the mean ± one standard deviation, range, and sample size (see text for explanation of variable abbreviations).

	P. blainvilii n = 65	*P. brasiliensis* n = 2	*P. dasythrix* n = 8	*P.* aff. *dasythrix* n = 8	*P. kerri* (holotype)	*P. lamarum* n = 23	*P. lundi* (holotype)
GSL	48 ± 1.7 (44.4 - 52.7)	50.4 ± 2.7 (48.5 - 52.3)	48.7 ± 2.0 (46.0 - 52.0)	48.8 ± 1.4 (47.1 - 50.7)	51.7	47.4 ± 2.0 (43.1 - 51.3)	47.7
NL	13.7 ± 0.9 (11.9 - 16.6)	14.2 ± 0.6 (13.8 - 14.6)	14.6 ± 1.2 (12.6 - 16.1)	14.9 ± 0.8 (13.6 - 16.2)	16.4	14.4 ± 1.1 (12.6 - 16.0)	13.4
RL	16.8 ± 0.9 (15.0 - 18.9)	18.6 ± 1.6 (17.5 - 19.8)	17.2 ± 2.1 (12.7 - 19.1)	17.4 ± 0.8 (16.3 - 18.6)	19.6	16.6 ± 1.0 (14.5 - 18.2)	16.6
OL	13.7 ± 0.5 (12.7 - 14.7)	13.8 ± 0.7 (13.3 - 14.3)	13.7 ± 0.5 (12.9 - 14.8)	13.7 ± 0.6 (12.9 - 14.8)	13.9	13.2 ± 0.6 (12.0 - 14.2)	13.2
RB	6.9 ± 0.3 (6.2 - 7.8)	7.3 ± 0.4 (7.0 - 7.6)	7.3 ± 0.4 (6.9 - 7.8)	7.2 ± 0.2 (7.0 - 7.4)	8.4	7.1 ± 0.5 (6.2 - 8.2)	6.7
IOC	10.1 ± 0.6 (8.8 - 11.9)	11.8 ± 0.9 (11.1 - 12.5)	10.4 ± 0.5 (9.9 - 11.3)	10.1 ± 0.5 (9.3 - 10.6)	12.1	10.8 ± 0.6 (9.9 - 12.3)	11.1
MB	18.8 ± 0.7 (17.6 - 20.5)	19.8 ± 1.9 (18.5 ± 21.2)	19.1 ± 0.5 (18.2 - 20)	19.4 ± 0.7 (18.7 - 20.6)	19.1	18.8 ± 0.7 (17.5 - 20.6)	17.9
ZB	23 ± 0.9 (20.9 - 25.3)	24.0 ± 2.0 (22.6 - 25.4)	23.0 ± 0.6 (22.4 - 23.7)	23.1 ± 0.5 (22.3 - 23.9)	23.3	22.6 ± 1.0 (20.1 - 24.1)	22.5
CIL	43.1 ± 1.6 (40.2 - 47.3)	44.8 ± 4.0 (42.0 - 47.6)	43.9 ± 1.7 (41.4 - 46.6)	43.4 ± 0.9 (42.4 - 45.1)	46.1	42.3 ± 1.9 (38.2 - 46.3)	41.9
BaL	37 ± 1.5 (33.4 - 40.4)	39.1 ± 4.3 (36.0 - 42.1)	37.7 ± 1.8 (35.2 - 40.0)	37.3 ± 1.1 (35.9 - 39.2)	39.8	36.2 ± 2.2 (32.5 - 40.6)	36.0
D	9.6 ± 0.7 (8.1 - 11.3)	10.7 ± 0.8 (10.1 - 11.2)	10.4 ± 0.7 (9.4 - 11.4)	10.1 ± 0.6 (9.5 - 10.8)	12.1	9.7 ± 0.7 (8.6 - 11.3)	10.1
MTRL	11.4 ± 0.5 (10.4 - 12.7)	11.8 ± 0.0 11.8	12.0 ± 0.5 (11.1 - 12.6)	11.5 ± 0.7 (10.7 - 12.6)	11.2	11.1 ± 0.5 (9.9 - 12.0)	11.0
PLb	8.5 ± 0.5 (7 - 9.6)	9.8 ± 0.5 (9.5 - 10.2)	11.2 ± 1.7 (9.2 - 13.1)	9.8 ± 1.7 (8.0 - 12.5)	8.3	8.2 ± 0.7 (7.0 - 9.5)	8.6
IFL	3.8 ± 0.3 (3 - 4.6)	4.8 ± 0.1 (4.7 - 4.8)	3.5 ± 0.4 (2.9 - 4.2)	3.8 ± 0.4 (3.3 - 4.5)	4.7	3.9 ± 0.3 (3.4 - 4.2)	4.7
BuL	10.2 ± 0.6 (9.1 - 11.8)	10.5 ± 0.6 (10.1 - 10.9)	9.5 ± 0.2 (9.1 - 9.7)	9.7 ± 0.3 (9.5 - 10.4)	9.9	10.2 ± 0.5 (9.3 - 11.4)	8.9
Pla	18.1 ± 0.9 (15.6 - 20.4)	20.2 ± 1.7 (21.4 - 22.8)	19.7 ± 1.1 (18.6 - 21.4)	18.9 ± 0.8 (17.7 - 20.0)	20.4	17.8 ± 1.1 (15.7 - 19.7)	17.8
PPL	19.6 ± 0.8 (17.4 - 21.5)	22.1 ± 1.0 (21.4 - 22.8)	19.1 ± 0.9 (18.0 - 20.3)	19.6 ± 0.4 (18.8 - 20.1)	20.7	20.0 ± 0.9 (18.3 - 21.9)	19.7
MPF	3.2 ± 0.2 (2.7 - 3.6)	4.0 ± 0.7 (3.5 - 4.5)	3.2 ± 0.2 (3.0 - 3.6)	3.5 ± 0.3 (3.1 - 3.9)	3.5	3.2 ± 0.3 (2.9 - 3.8)	3.1
MaxB	7.5 ± 0.3 (6.4 - 8.2)	8.3 ± 0.1 (8.2 - 8.4)	7.9 ± 0.4 (7.1 - 8.2)	7.6 ± 0.7 (6.8 - 8.8)	8.1	7.4 ± 0.5 (6.4 - 8.3)	7.4
OccW	8.3 ± 0.2 (7.9 - 9.0)	8.5 ± 0.1 8.4 - 8.6	7.9 ± 0.3 (7.6 - 8.5)	8.6 ± 0.6 (7.8 - 9.2)	8.6	8.4 ± 0.2 (8.1 - 8.9)	8.7
RD	10.3 ± 0.4 (9.1 - 11.1)	11.1 ± 1.4 (10.1 - 12.1)	10.6 ± 0.8 (9.3 - 12.0)	10.5 ± 0.8 (9.1 - 11.5)	11.3	10.0 ± 0.7 (8.6 - 11.4)	10.4
CDM1	13.7 ± 0.5 (12.8 - 14.9)	14.4 ± 1.1 (13.7 - 15.2)	13.9 ± 0.5 (13.0 - 14.8)	13.8 ± 0.5 (13.2 - 14.5)	14.4	13.5 ± 0.6 (12.2 - 14.8)	13.6
CD	18.5 ± 0.7 (16.5 - 19.9)	17.5 ± 3.5 (15.0 - 20.0)	18.8 ± 1.3 (16.7 - 20.3)	19.2 ± 1.1 (17.2 - 20.2)	19.4	17.8 ± 1.0 (15.8 - 19.1)	17.8

TABLE 6 (continued)

	P. mantiqueirensis (holotype)	P. medius n = 10	P. nigrispinus n = 20	P. pattoni n = 30	P. thomasi n = 16	P. unicolor (holotype)
GSL	48.1	52.5 ± 3.5 (46.2 - 56.7)	49.8 ± 2.6 (45.6 - 55.5)	49.2 ± 2.5 (44.7 - 54.6)	60.2 ± 2.6 (56.2 - 65.7)	-
NL	14.2	15.6 ± 1.8 (13 - 18.1)	15.0 ± 1.7 (12.2 - 17.9)	14.7 ± 1.3 (11.8 - 17.6)	18.9 ± 1.7 (15.6 - 21.9)	-
RL	17.2	19.0 ± 1.5 (15.9 - 20.7)	17.8 ± 1.4 (15.6 - 20.2)	17.3 ± 1.3 (15.2 - 20.8)	22.4 ± 1.9 (19.6 - 25.9)	-
OL	14.0	14.4 ± 1 (12.8 - 16.5)	13.9 ± 0.6 (12.6 - 14.7)	13.5 ± 0.5 (12.2 - 14.3)	16.0 ± 0.7 (14.9 - 17.4)	-
RB	7.9	7.8 ± 0.6 (7.0 - 8.7)	7.5 ± 0.5 (6.5 - 8.4)	7.8 ± 0.7 (6.5 - 9.0)	10.0 ± 0.6 (9.2 - 11.4)	9.6
IOC	10.2	11.1 ± 0.8 (10.3 - 13.1)	10.6 ± 0.5 (9.9 - 11.5)	11.4 ± 1.0 (9.0 - 13.4)	13.3 ± 0.6 (12.5 - 14.2)	-
MB	18.9	20.1 ± 0.9 (18.8 - 21.4)	19.7 ± 0.9 (18.1 - 21.1)	20.1 ± 1.1 (17.7 - 22.1)	22.5 ± 0.9 (21.0 - 24.4)	22.7
ZB	23.1	24.5 ± 1.5 (21.6 - 26.8)	24.1 ± 1.1 (22.1 - 26.3)	24 ± 1.2 (21.2 - 26.4)	28.5 ± 0.9 (26.8 - 29.8)	27.7
CIL	43.0	47.5 ± 3.4 (40.8 - 51.4)	44.3 ± 2.5 (40.3 - 49.8)	43.6 ± 2.4 (39.6 - 48.4)	53.7 ± 2.4 (50.6 - 59.3)	-
BaL	36.3	40.8 ± 3.3 (34.6 - 44.5)	38.4 ± 2.1 (33.4 - 42.4)	37.6 ± 2.4 (33.2 - 42.0)	46.2 ± 2.3 (43.7 - 51.0)	44.9
D	10.0	11.3 ± 1.1 (8.6 - 12.6)	10.5 ± 0.9 (8.7 - 12.6)	10.3 ± 0.8 (8.6 - 12.1)	13.4 ± 1.1 (11.6 - 15.5)	11.2
MTRL	11.4	12.5 ± 1.0 (11.0 - 14.0)	11.7 ± 0.5 (11.0 - 13.0)	11.0 ± 0.6 (9.7 - 12.1)	13.9 ± 0.6 (12.9 - 14.9)	13.9
PLb	9.0	9.8 ± 0.7 (8.7 - 10.5)	9.2 ± 0.6 (8.3 - 10.5)	8.7 ± 0.8 (6.9 - 9.9)	10.7 ± 0.7 (9.7 - 11.8)	12.2
IFL	3.9	3.8 ± 0.5 (3.4 - 4.9)	3.6 ± 0.6 (3.0 - 4.9)	4.2 ± 0.5 (3.3 - 5.4)	5.2 ± 0.6 (4.4 - 6.1)	4.6
BuL	9.4	9.7 ± 0.6 (8.7 - 10.7)	10.0 ± 0.7 (8.2 - 11.2)	10.3 ± 0.7 (8.7 - 11.8)	11.2 ± 0.6 (10.4 - 12.2)	12.8
Pla	18.9	21.0 ± 1.7 (17.8 - 23.2)	19.5 ± 1.2 (17.6 - 22.9)	19.0 ± 1.4 (16.2 - 21.6)	24.1 ± 1.4 (22.2 - 27.0)	23.1
PPL	18.7	21.1 ± 1.4 (19.3 - 22.9)	19.9 ± 1.1 (17.7 - 22.5)	20.1 ± 1.2 (17.4 - 22.1)	23.2 ± 1.2 (21.1 - 25.1)	-
MPF	3.1	3.7 ± 0.4 (3.3 - 4.7)	3.4 ± 0.4 (2.4 - 4.0)	3.5 ± 0.3 (2.9 - 4.1)	4.5 ± 0.3 (4.0 - 5.0)	3.5
MaxB	8.3	8.1 ± 0.5 (7.5 - 9.0)	7.7 ± 0.5 (7.0 - 8.9)	7.5 ± 0.4 (6.8 - 8.2)	9.4 ± 0.4 (8.4 - 10.1)	8.8
OccW	8.6	8.8 ± 0.6 (8.2 - 9.9)	8.9 ± 0.5 (8.0 - 10.0)	8.8 ± 0.4 (8.0 - 9.6)	10.0 ± 0.5 (9.2 - 11.1)	-
RD	9.9	11.1 ± 1.0 (9.8 - 12.7)	10.7 ± 0.7 (8.9 - 12.0)	10.4 ± 0.7 (9.1 - 12.1)	13.4 ± 0.6 (12.4 - 14.5)	12.4
CDM1	13.9	15.0 ± 1.3 (13.4 - 17.3)	14.3 ± 0.6 (13.6 - 15.8)	14.0 ± 0.7 (12.6 - 15.0)	17.6 ± 0.7 (16.5 - 18.9)	-
CD	19.7	18.6 ± 1.2 (16.5 - 20.0)	19.9 ± 1.1 (17.8 - 21.6)	19.0 ± 1.1 (16.8 - 21.1)	20.3 ± 1.3 (18.8 - 21.7)	-

TABLE 7. Principal component eigenvalues for log transformed cranial variables of *Phyllomys* spp. when all 13 species are included.

Variable	PC-1	PC-2	PC-3	PC-4	PC-5
Log RB	0.918	0.069	-0.123	-0.031	-0.022
Log MB	0.896	0.250	-0.108	0.033	0.072
Log D	0.900	0.075	-0.131	-0.280	-0.203
Log MTRL	0.751	-0.457	0.372	-0.097	0.222
Log BuL	0.565	0.673	0.445	0.143	-0.061
Log Pla	0.941	-0.081	-0.028	-0.251	-0.101
Log MPF	0.832	0.172	-0.342	0.236	0.236
Log MaxB	0.744	-0.440	0.032	0.440	-0.241
Log RD	0.903	-0.147	0.093	-0.048	0.092
Eigenvalue	6.284	0.986	0.507	0.424	0.232
% contribution	69.8	11.0	5.6	4.7	2.6

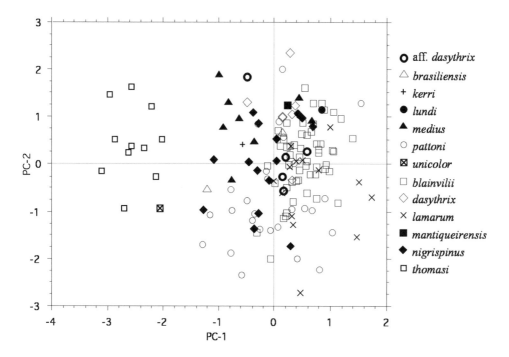

FIGURE 32. Bivariate plot of the first versus second principal component axes for adult individuals of *Phyllomys* spp. of both sexes and from all localities.

The next set of analyses focused on closely related taxa and species that show very similar morphologies and are therefore hard to tell apart. Figure 33 shows a ratio diagram comparing members of the northeastern clade (*P. blainvilii, P. brasiliensis, P. lamarum*) and *P. pattoni*, which has been historically confused with *P. brasiliensis* (see Emmons et al., 2002). *Phyllomys brasiliensis* is a larger animal, as reflected in nearly all measurements, notably the incisive foramen length (IFL) and mesopterygoid fossa width (MPF). *Phyllomys pattoni* is the second largest, having a distinctively wide skull (RB, MB, ZB, OccW) but the shortest toothrow on average (MTRL). *Phyllomys blainvilii* and *P. lamarum* have similar measurements, but the latter has longer nasals (NL), a broader rostrum (RB), wider interorbital region (IOC), and shorter toothrow (MTRL) and palate (PLb) than the former. They are well separated in the discriminant analysis as shown in the scatterplot of canonical scores for the two functions extracted (Fig. 34). The first discriminant function accounts for 87% of the variation (Table 8) and separates *P. lamarum* from *P. blainvilii*, with most of the variation due to general skull size, especially the rostrum (GSL, NL, RL, D, PLa, RD).

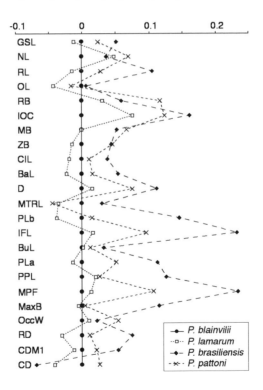

FIGURE 33. Ratio diagram comparing the mean $\log_{10}$ values for each cranial dimension among *P. blainvilii, P. lamarum, P. brasiliensis,* and *P. pattoni.*

TABLE 8. Standardized discriminant function coefficients for log-transformed cranial variables of *P. blainvilii*, *P. lamarum*, and *P. pattoni*.

Variables	DF-1	DF-2
Log GSL	2.08845	0.428959
Log NL	0.89113	-0.478981
Log RL	-1.41135	0.902901
Log OL	-0.60857	0.549245
Log RB	0.23695	0.147798
Log IOC	0.57325	-0.489563
Log MB	0.53185	0.678397
Log ZB	-0.26562	-0.089424
Log CIL	-0.48077	0.690209
Log BaL	-0.60592	-0.710774
Log D	1.06772	0.284469
Log MTRL	-0.43538	-0.206334
Log PLb	0.51993	-0.017267
Log BuL	-0.67756	-0.371568
Log PLa	-0.79298	-0.903644
Log PPL	-0.25779	-0.321012
Log MPF	0.23324	0.422405
Log MaxB	0.23667	0.018668
Log RD	-0.89220	-0.527454
Log CDM1	0.36068	0.354301
Eigenvalue	4.01762	0.600104
% contribution	87.004	12.996

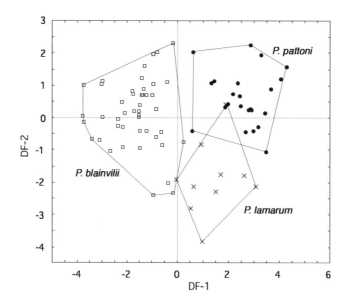

FIGURE 34. Bivariate plot of discriminant scores for the first two axes in comparisons between adult *P. blainvilii*, *P. lamarum*, and *P. pattoni*, based on $\log_{10}$ cranial variables.

Comparisons between *P. nigrispinus*, *P. kerri*, and *P. thomasi* show obvious size differences, as *P. thomasi* is the largest species in the genus (Fig. 35). This is again reflected in the principal components analysis, where the first axis accounts for 79% of the variation, and almost all variables contribute equally highly (Table 9). Unfortunately there is only one individual of *P. kerri* in toothwear age category 8 (MNRJ 6241, the holotype), impeding assessment of variation within the group. *Phyllomys kerri* is proportionally smaller than *P. thomasi* in all measurements, and it differs from *P. nigrispinus* mainly by having longer and wider rostrum (NL, RL, RB), wider interorbital region (IOC), longer diastema (D) and incisive foramen (IFL), and a shorter palate (PLb). In the scatterplot of the principal components, *P. kerri* is generally marginal to the cluster of *P. nigrispinus* (Fig. 36). More samples are necessary in order to evaluate the actual degree of morphological differentiation between those two species.

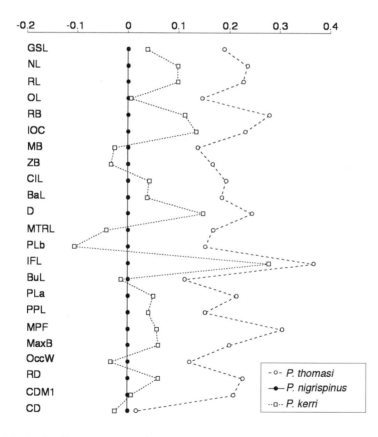

FIGURE 35. Ratio diagram comparing the mean $\log_{10}$ values for each cranial dimension among *P. thomasi*, *P. nigrispinus*, and *P. kerri*.

TABLE 9. Principal component eigenvalues for log-transformed cranial variables of *Phyllomys nigripinus*, *P. kerri*, and *P. thomasi*.

Variables	PC-1	PC-2	PC-3	PC-4	PC-5
Log GSL	0.979	0.001	-0.117	-0.053	-0.049
Log NL	0.794	0.547	0.014	0.081	-0.041
Log RL	0.885	0.377	0.021	0.066	-0.190
Log OL	0.903	-0.117	-0.160	-0.069	0.026
Log RB	0.922	0.001	0.137	0.076	-0.116
Log IOC	0.924	-0.083	-0.103	-0.112	-0.221
Log MB	0.874	-0.164	0.035	-0.111	0.328
Log ZB	0.946	0.055	0.107	-0.001	0.103
Log CIL	0.980	0.037	-0.084	-0.059	-0.004
Log D	0.887	0.277	-0.192	-0.258	-0.079
Log MTRL	0.907	-0.214	0.094	0.279	-0.005
Log PLb	0.800	-0.243	-0.005	0.512	0.016
Log BuL	0.806	0.085	-0.347	0.041	0.438
Log PLa	0.975	0.047	-0.090	0.017	-0.077
Log MPF	0.827	0.127	0.400	-0.137	0.233
Log MaxB	0.797	-0.211	0.457	-0.178	-0.033
Log Occw	0.689	-0.585	-0.167	-0.232	-0.121
Log CDM1	0.966	-0.107	-0.070	0.082	-0.116
Log RD	0.965	0.065	0.097	0.034	-0.033
Eigenvalue	15.016	1.103	0.675	0.573	0.511
% contribution	79.0	5.8	3.6	3.0	2.7

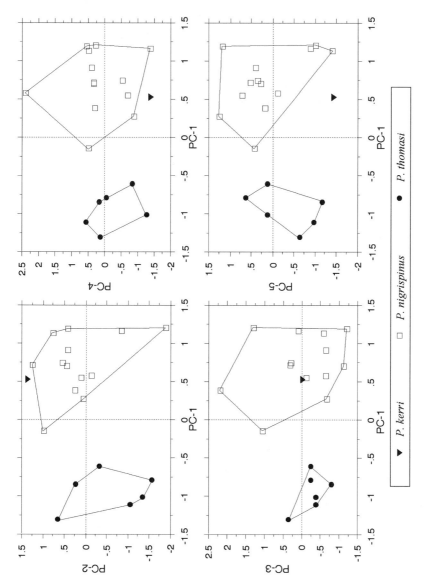

FIGURE 36. Bivariate plots of the first versus second (lower left), third (lower right), fourth (upper left), and fifth (upper right) principal component axes for adult individuals of *P. kerri*, *P. nigrispinus*, and *P. thomasi* of both sexes and from all localities, based on log₁₀ cranial variables.

Phyllomys medius is larger than *P. dasythrix* and *P.* aff. *dasythrix* in all measurements, except PLb and CD (Fig. 37). *Phyllomys* aff. *dasythrix* has smaller palate (PLb) and wider mesopterygoid fossa (MPF) than *P. dasythrix.* In the discriminant analysis, *P.* aff. *dasythrix* differentiates from the other two on the first axis, which is influenced by the skull length (GSL), interorbital constriction (IOC), and the toothrow length (MTRL), while *P. medius* and *P. dasythrix* separate on the second axis, which is affected mainly by another measure of skull length (CIL) and the orbital length (see Table 10 and Fig. 38).

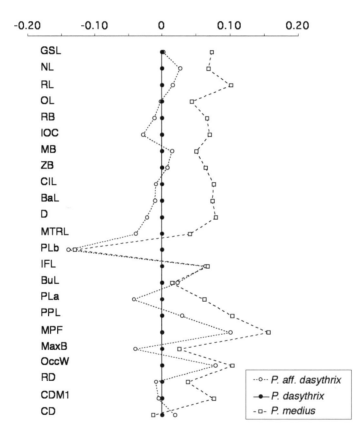

FIGURE 37. Ratio diagram comparing the mean $\log_{10}$ values for each cranial dimension among *P.* aff. *dasythrix*, *P. dasythrix*, and *P. medius*.

TABLE 10. Standardized discriminant function coefficients for log-transformed cranial variables of *P. dastyrhix*, *P.* aff. *dasythrix*, and *P. medius*.

Variable	DF-1	DF-2
Log GSL	5.79655	0.13798
Log OL	0.26287	-1.11676
Log IOC	-3.22173	-0.17260
Log MB	-0.78051	-0.20017
Log CIL	-1.63039	1.29975
Log MPF	1.75534	0.56336
Log D	1.76263	0.08395
Log MTRL	-2.33104	-0.27826
Log PLB	-0.18621	-0.69961
Log PPL	-1.34361	0.77987
Eigenvalue	6.71680	2.30477
% contribution	74.45	25.55

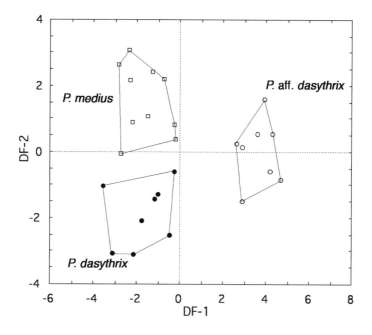

FIGURE 38. Bivariate plot of discriminant scores for the first two axes in comparisons between adult *P.* aff. *dasythrix*, *P. dasythrix*, and *P. medius* based on $\log_{10}$ cranial variables.

NATURAL HISTORY

In this section I report on general aspects of the natural history and ecology of the Atlantic tree rats as deduced from the fragmentary data. Because the number of observations is too small for any statistical treatment, I describe only the data available on reproduction, diet, arboreal habits, and abundance.

Reproduction

Very little is known regarding reproductive cycles. Table 11 presents a compilation of data from specimen tags and a few published accounts (Burmeister, 1854; Davis, 1945; Moojen, 1952; Reinhardt, 1849). The number of embryos is usually one or two, and pregnant females are found mainly during the rainy season, from September through May in nearly all species and localities. In *P. blainvilii*, Moojen (1952) reported up to four embryos, but he mentioned that one was the most common litter size. He also pointed out that young and young-adult animals were trapped in the same period (February and March), suggesting that they were born in August–September. Only one female (YL 200, *P. pattoni*) collected in April was recorded as lactating. Table 12 shows when young individuals (toothwear age categories 1–4) were collected at different localities. Young are likewise collected mainly during the rainy season (October–May), suggesting that reproduction probably starts towards the end of the dry season as hypothesized by Moojen (1952). This is a general trend observed, for example, at Terezópolis (Davis, 1945), where most species of rodents begin to breed in July and August and reach full reproductive activity in October, when young become significantly more abundant. In this cycle, females are pregnant and/or nursing during the rainy season, when resources are more abundant, maximizing reproductive effort and survival of young.

A sternal gland is often present on the upper chest of adult individuals of both sexes, reaching up to 25 x 10 mm in size. Externally, the skin appears naked and thin, and a secretion that exudes a weak odor stains the fur. The stain on the chest of museum skins is probably indicative of an active gland. This stain was present on 11 museum skins of *P. blainvilii*, 6 males and 5 females (2 of them pregnant), all trapped between November and March. Among *P. lamarum*, 3 males and 4 females had active glands; in *P. pattoni*, 2 males trapped in February and August, in *P.* aff. *dasythrix* one male and one female trapped in March, and in *P. lundii* one male (the holotype) trapped in October. As mentioned by Olmos (1997), this gland is probably associated with scent-marking and may play a role in territoriality in this genus as in other arboreal folivore echmyids, such as the bamboo rats *Dactylomys* and *Kannabateomys* (Emmons, 1981; Silva, 1993).

TABLE 11. Reproductive data on pregnant females of *Phyllomys* compiled from the literature and museum tags.

Species	Number of embryos	Month	Locality	Source
P. blainvilii	1	Oct	Crato, CE	Moojen (1952); skin tag (MZUSP 6147)
P. blainvilii	1–4	Feb–Mar	Bom Jesus da Lapa, BA	Moojen (1952, as "*P. lamarum*")
P. brasiliensis	1	Sep	Lagoa Santa, MG	Reinhardt (1849); skin tag (UZMC 81)
P. lamarum	2	May	Lamarão, BA	Skin tag (BMNH 3.9.5.96)
P. medius	2	Sep	Terezópolis, RJ	Davis (1945); skin tag (MNRJ 6240)
P. nigrispinus	1	Oct	Interlagos, SP	Skin tag (MZUSP 10311)
P. pattoni	2	Feb	Nova Friburgo, RJ	Burmeister (1854)
P. pattoni	2	Aug	Viçosa, MG	Skin tag (UFV 696)

TABLE 12. Young individuals (toothwear age categories 1–4) of *Phyllomys* recorded according to the month they were trapped at different localities.

Species	Locality	Age category	Month
P. blainvilii	Crato, CE	1	Nov
		3	Oct, Nov, Jan, Feb, Apr
		4	Jan
	Bom Jesus da Lapa, BA	3	Feb, Mar
	Seabra, BA	4	Nov
	Andaraí, BA	3	May
P. dasythrix	São Francisco de Paula, RS	4	Feb
P. nigrispinus	Ribeirão Fundo, SP	2	Jul
	Ilha do Cardoso, SP	3	Nov
	Interlagos, SP	3	Oct
	São Paulo, SP	3	Oct
	Taboão da Serra, SP	3	Apr
P. pattoni	Ilhéus, BA	1	Mar

Gut Morphology and Diet

The general morphology of the digestive tract of *Phyllomys* is similar to that of other arboreal echimyids (Emmons, 1981; Silva, 1993). These rodents are primarily herbivores as indicated by their very long and extensively folded intestines and well-developed caecum. The examination of the stomach contents of one specimen of *P. pattoni* also points to a herbivorous diet, as it was filled with a compact green-yellowish mass of plant material. The caecum and large intestine of *P. pattoni* are relatively larger than in *Mesomys* and *Kannabateomys* but smaller than in *Dactylomys* (Table 13). The small intestine of *Phyllomys* is the longest found among these genera, indicating the possibility of a more diverse diet like *Mesomys*, which includes fruits, leaves, and insects.

TABLE 13. Mean relative size (in percent of head and body length) of segments of the digestive tracts of arboreal echimyids. Measurements were taken from specimens fixed in formalin and kept in 70% ethanol, except for *K. amblyonyx*, which were taken immediately after death.

	Species			
	Phyllomys pattoni (n = 1)	*Mesomys hispidus* (n = 2)	*Dactylomys dactylinus* (n = 2)	*Kannabateomys amblyonyx* (n = 3)
Head and body length (mm)	230	188	290	273
Small intestine:				
Length	370	344	159	174
Maximum width	2	3	5	—
Large intestine:				
Length	226	140	247	172
Maximum width	4	2	7	—
Caecum:				
Length	57	48	66	37
Maximum width	9	8	13	—
Source	Present report	Emmons (1981)	Emmons (1981)	Silva (1993)

Arboreal Habits and Abundance

Like all members of the subfamily Echimyinae, Atlantic tree rats are adapted to arboreal locomotion. They have relatively short limbs, short and broad feet with strong claws, long tails, and long vibrissae. This is confirmed by the data on skin tags: out of 115 captures of *Phyllomys* (pooling all species together) where the approximate height of capture was recorded, only three were on the ground, and the remaining were in trees between 1 and 20 meters. Of the 141 cases where the method of capture was reported on skin tags, 53 animals were shot, 30 were live-trapped, 30 were beaten with sticks, and 28 were collected with bare hands. As pointed out by Voss and Emmons (1996) collecting arboreal echimyids requires a substantial commitment of time and effort, and night hunting is the most effective method. Looking for nests in tree hollows during the day is also effective, considering that 52 of the specimens mentioned above were captured in that manner. Another successful way of obtaining large series of specimens is by catching them by hand or with nets from a boat when areas are flooded subsequent to dam construction.

Specimens of the Atlantic tree rats are rare in museum collections. This raises the question of whether they are rare in nature, difficult to collect, or both. Even studies sampling the arboreal environment, including extensive canopy trapping, failed to obtain series of *Phyllomys* (e.g., Davis, 1945; Fonseca and Kierulff, 1989; Laemmert et al., 1946), usually producing one to a few specimens after tens of thousands of trap nights. At the same time these studies were successful in trapping other arboreal small mammals, such as the marsupials *Caluromys*, *Micoureus*, and *Gracilinanus* and the sigmodontine rodents *Rhipidomys* and *Oecomys*. Studies employing alternative methods of capture, such as hunting and tree hollow probing, are much more successful in obtaining specimens. During the plague survey in northeastern Brazil, for example, 120 specimens of *P. blainvilii* were collected in the vicinity of Crato, CE, between 1952 and 1953 (Freitas, 1957). João Moojen collected 15 specimens of *P. blainvilii* near Bom Jesus da Lapa, BA, during one week; A. M. Olalla collected 12 *P. nigrispinus* in less than two weeks at Interlagos, SP; and I collected seven *P. pattoni* in three days at Caravelas, BA. These results indicate that some species of *Phyllomys* may be locally abundant, but unconventional trapping methods must be used to collect them. This is true not only for *Phyllomys*, but for arboreal echimyids including tree rats and bamboo rats (e.g., Patton et al., 2000; see also Voss and Emmons, 1996; Voss et al., 2001), which are folivorous and do not seem to be attracted by the different baits used in live traps, as also suggested by Patton et al. (2000).

CONSERVATION

Conservationists largely ignore rodents, despite the fact that this group encompasses nearly half of the mammal species of the world (Amori and Gippoliti, 2001). The fact that *Phyllomys* is endemic to the Atlantic forest (*sensu lato*) of Brazil is by itself of conservation interest since most of the original forest cover is gone (see Conservation International do Brasil et al., 2000; Fonseca, 1985; Myers et al., 2000). The scarcity of museum specimens, and consequent lack of knowledge about nearly all species, and the limited known distribution of some species (see Figs. 26, 27, 28, 29, 30) aggravate this neglect. Some taxa are known only from the holotype, either collected long ago, like *P. unicolor*, or recently, like *P. mantiqueirensis*. Others are represented by very few specimens, from small geographic areas, such as *P. lundi* (two specimens), *P. kerri* (three specimens), and *P. brasiliensis* (five specimens), and most species are known from fewer than 30 museum records from a small number of localities: *P.* aff. *dasythrix* (13 specimens from 9 localities), *P. dasythrix* (16 specimens from 7 localities), *P. thomasi* (22 specimens from 3 localities on an island), and *P. medius* (28 specimens from 15 localities).

Ignorance regarding the systematic status of the genus *Phyllomys* and its contents has led researchers to misidentify several species over the years, with critical implications for their conservation status. For example, the recently described *P. pattoni* is a taxon that has been treated as "*P. brasiliensis*" for over a century (Emmons et al., 2002). Compared to *P. brasiliensis*, *P. pattoni* has an extensive geographic distribution, is relatively well represented in museum collections, has frequently been recorded at several localities, and therefore does not seem to be under threat. *Phyllomys brasiliensis*, on the other hand, was until very recently known only from a few animals collected more than 150 years ago. This contrast shows the fundamental importance of systematic and taxonomic work in conservation. If one is not able to correctly identify a taxon, nothing can be done to protect it, or protection measures (with the associated costs) may be included in plans for taxa actually less threatened than others.

Another important conservation issue is the large gap in our knowledge of Neotropical mammal diversity, especially in South America (Patton et al., 1997; Vivo, 1996; Voss and Emmons, 1996). For example, the number of known lowland rainforest species increased by more than 10% between 1990 and 1997 (Emmons and Feer, 1997). Although most of this increase is due to the elevation of taxa formerly treated as subspecies (e.g., Lara and Patton, 2000; Mustrangi and Patton, 1997; Patton and Silva, 1997), several species or even genera had never been recorded before. For example, more than 10% of the species of mammals collected in a recent study in the Amazon basin were new to science (Patton et al., 2000). However, one does not have to go to the farthest corner of the world to find new taxa. In the present report, I describe two new species, *P. lundi* and *P.*

mantiqueirensis, that are found in southeastern Brazil, between São Paulo, Rio de Janeiro, and Belo Horizonte, the largest metropolitan areas in the country. In addition, Rio de Janeiro and São Paulo house two of the largest mammal collections in South America, and have historically concentrated most of the mammalogists who have engaged in the majority of collecting efforts in the country. There is still much to be explored, discovered, and documented regarding Neotropical mammal diversity, even in the most densely populated and best-studied areas of Brazil.

Phylogenetic analyses are shedding light on the ecological and evolutionary processes that have shaped current biodiversity (Purvis and Hector, 2000). Phylogenetic data should be carefully considered in defining priorities for conservation, although this is rarely done (e.g., Silva and Patton, 1998). If we want to preserve not only species but also their history and genetic diversity, we must take into account results from systematic and phylogeographic studies (see Vane-Wright et al., 1991). Regarding the genus *Phyllomys,* the region in southeastern Brazil along the Serra do Mar and Serra da Mantiqueira is the most important, since several species occur within this small geographic area (Fig. 17). Not only are several species found there, but most known phylogenetic lineages are represented in that area, including old branches, such as that leading to *P. mantiqueirensis.* Indeed this region was considered of "extreme biological importance" in a recent workshop for defining conservation priorities in the Atlantic forest (Conservation International do Brasil et al., 2000).

Regionally, *Phyllomys brasiliensis* is listed as threatened in the state of Minas Gerais (as "*Echimys (Nelomys) braziliensis,*" Fonseca and Sábato, 1998), and *P. thomasi* is listed in the state of São Paulo (as "*Nelomys thomasi,*" State Decree no. 42838, 4 February 1998). In global terms, the 2000 IUCN Red List of Threatened Species (Hilton-Taylor, 2000) records two species of *Phyllomys* (as *Echimys,* following Woods, 1993), both of them assessed in 1996: *P. blainvilii* is listed as at "lower risk, near threatened" and *P. thomasi* as "vulnerable." The major threat to both of them is "human settlement."

Using the revised IUCN categories and corresponding criteria for their application (version 3.1, IUCN, 2001) and based on our current knowledge, I suggest the following categorization for certain species of *Phyllomys:*

- Critically Endangered (CR)—*P. unicolor* and *P. mantiqueirensis.*
- Endangered (EN)—*P. brasiliensis, P. lundi,* and *P. thomasi.*
- Least Concern (LC)—*P. pattoni.*
- Data Deficient (DD)—*P. blainvilii, P. dasythrix, P. medius, P. lamarum,* and *P. nigrispinus.*

Phyllomys kerri is known from three specimens only, but it is not listed above since its species status is still questionable (see remarks under the species account above).

The holotype and only known specimen of *P. unicolor* was collected prior to 1824, suggesting the possibility that this species may actually be extinct. However, I believe that it remains extant, based on fieldwork carried out in 1998 by L. P. Costa and myself at Caravelas, BA, near the type locality of *P. unicolor*. After helping us to collect a series of *P. pattoni*, two locals mentioned the existence of another species of tree rat in the area. According to them, this rat is similar but larger, darker, with soft pelage, all general external attributes that match the description of *P. unicolor*. The only other large tree rat that may occur in this area is the painted tree rat, *Callistomys pictus*, which has bold external markings precluding confusion with any other species. Intensive field studies should be carried out at Caravelas to confirm the existence and status of *P. unicolor*. Until then, this species should be considered Critically Endangered (CR B1ab(iii), IUCN, 2001) because its extent of occurrence is estimated to be less than 100 km^2, it is known to exist at only a single location, and there has been continuing decline in the area and quality of habitat.

Phyllomys mantiqueirensis is known only from a single area that belongs to the Brazilian Army and is therefore currently protected from indiscriminate exploitation. However, it may be restricted to high-elevation forests in the Serra da Mantiqueira, a very specific and naturally restricted habitat. The locality is within the limits of the Área de Proteção Ambiental (APA) da Mantiqueira. This type of protected area allows the use of the natural resources but in a restricted and controlled way. *Phyllomys mantiqueirensis* should be listed as Critically Endangered (CR B1ab(iii), IUCN, 2001) because its extent of occurrence is estimated to be less than 100 km^2, it is known to exist at only a single location, and there has been continuing decline in the area and quality of habitat.

Phyllomys brasiliensis was recorded recently from private lands, but its historic range includes what is now the APA Carste de Lagoa Santa, MG (ca. 36,000 ha), created in 1990 by IBAMA (Instituto Brasileiro do Meio Ambiente e dos Recursos Naturais Renováveis) to protect the impressive number of limestone caves and the associated biota of the region. *Phyllomys brasiliensis* should be listed as Endangered (EN B1ab(iii), IUCN, 2001) because its extent of occurrence is estimated to be less than 5000 km^2, it is known to exist at no more than five locations, and there has been continuing decline in the area and quality of habitat.

The type locality of *P. lundi* is in a private unprotected area, but this species also occurs at Poço das Antas, RJ (ca. 5000 ha), a federal biological reserve managed by IBAMA, and created to protect the golden lion tamarins (*Leontopithecus rosalia*). *Phyllomys lundi* should be considered Endangered (EN B1ab(iii), IUCN, 2001) because its extent of occurrence is estimated to be less than

5,000 km², it is known to exist at no more than five locations, and there has been continuing decline in the area and quality of habitat.

Phyllomys thomasi is endemic to the island of São Sebastião, where 80% of the 336 km² consist of Atlantic rainforest, mostly old second growth due to selective logging. The forest is, however, currently in the process of expansion (Olmos, 1997) since most of the island surface is protected as a state park (Parque Estadual de Ilhabela, SP, ca. 27,000 ha). *Phyllomys thomasi* should be listed as Endangered (EN B2ab(iii), IUCN, 2001) because its area of occupancy is less than 500 km², it is known to exist at no more than five locations, and there has been continuing decline in the quality of habitat.

The remaining species are better represented in museum collections and have larger geographic ranges. *Phyllomys pattoni* has been frequently collected thoughout a broad range, including several large protected areas and is therefore of Least Concern (IUCN, 2001). The information available is inadequate to make a direct, or indirect, assessment of the risk of extinction of the other five species, and they should therefore be listed as Data Deficient (IUCN, 2001). *Phyllomys blainvilii* is protected at the Floresta Nacional do Araripe, CE (ca. 38,000 ha), which is inserted in the APA Chapada do Araripe. *Phyllomys dasythrix* occurs at the Parque Estadual de Itapuã, RS, a state park. *Phyllomys medius* is found at Reserva Natural Salto Morato, PR, a privately owned protected area (Reserva Particular do Patrimônio Natural—RPPN) recognized by IBAMA and inserted in the APA de Guaraqueçaba. *Phyllomys lamarum* occurs at Estação Ecológica de Acauã, MG (ca. 5200 ha), a state park. The type locality of *Phyllomys nigrispinus* is now protected as a national forest, Floresta Nacional de Ipanema, SP (ca. 5200 ha), and it also occurs at the Parque Estadual da Ilha do Cardoso, SP (ca. 22,500 ha).

Besides *Phyllomys*, 14 other rodent genera are endemic to the Atlantic forest: *Abrawayaomys, Blarinomys, Brucepattersonius, Delomys, Juliomys, Phaenomys, Rhagomys, Thaptomys, Wilfredomys, Callistomys, Chaetomys, Euryzygomatomys, Kannabateomys,* and *Trinomys,* many of which are known only from a few specimens. The main direct action that will help to protect the Atlantic tree rats, and Atlantic forest rodents and other small mammals in general, is to increase our understanding of their systematics, population genetics, distribution, natural history, and ecology. Specimens and associated data collected and adequately preserved by well-trained biologists, and deposited in accredited institutions throughout the country, are the main source of such studies. Research projects aimed at these goals (or even general mammal inventories) must be encouraged if we want to expand our knowledge of these fascinating animals, even if these studies involve collecting museum specimens of potentially threatened species of *Phyllomys*. The main threat to the Atlantic tree rats comes from human-induced habitat loss and not direct overexploitation by humans. In addition, mammalogists are very ineffective collectors of tree rats, judging by the number of museum specimens available, while deforestation has been highly effective in destroying the Atlantic forest (see Dean, 1995). As properly pointed out by Vivo (1996, p.): "A large number of well intentioned, but misinformed, people work in State and Federal institutions devoted to protect the fauna, flora, and environment. They usually share with many biologists the view that we already know everything we need to know of the Brazilian mammals, and this makes them strongly oppose attempts to collect specimens." The results of the present report show how poorly known the Atlantic tree rats are. The results also reinforce the fact that museum specimens provide the best material for the study of natural populations, especially in conjunction with modern genetic tools. These types of studies are essential in recording and understanding biodiversity and should be prioritized and carried out as quickly as possible.

Literature Cited

Almeida, F. F. M.
  1976 The system of continental rifts bordering the Santos Basin, Brazil. Anais da Academia Brasileira de Ciências 48:15–26.

Amori, G., and S. Gippoliti
  2001 Identifying priority ecoregions for rodent conservation at the genus level. Oryx 35:158–165.

Auler, A. S., and P. L. Smart
  2001 Late Quaternary Paleoclimate in semiarid northeastern Brazil from U-series dating of travertine and water-table speleothems. Quaternary Research 55:159–167.

Behling, H., H. W. Arz, J. Pätzold, and G. Wefer
  2000 Late Quaternary vegetational and climate dynamics in northeastern Brazil, inferences from marine core GeoB 3104. Quaternary Science Reviews 19:981–994.

Bookstein, F. L., B. Chernoff, R. L. Elder, J. M. Humphries, Jr., G. R. Smith, and R. E. Strauss
  1985 Morphometrics in Evolutionary Biology: The Geometry of Size and Shape Change, with Examples from Fishes. Special Publication 15, The Academy of Natural Sciences of Philadelphia, Philadelphia. 277 pp.

Bremer, K.
  1988 The limits of amino-acid sequence data in angiosperm phylogenetic reconstruction. Evolution 42:795–803.

Brown, K. S.
  1987 Areas where humid tropical forest probably persisted. Pp. 44–45 in Biogeography and Quaternary History in Tropical America (T. C. Whitmore and G. T. Prance, eds.). Clarendon Press, Oxford.

Burmeister, H.
1854 Systematische Uebersicht der Thiere Brasiliens: welche während einer Reise durch die Provinzen von Rio de Janeiro und Minas geraës gesammlt oder beobachtet Wurden. G. Reimer, Berlin.

Bush, G. L., S. M. Case, A. C. Wilson, and J. L. Patton
1977 Rapid speciation and chromosomal evolution in mammals. Proceedings of the National Academy of Sciences 74:3942–3946.

Bush, M. B., P. A. Colinvaux, M. C. Wiemann, D. R. Piperno, and K. B. Liu
1990 Late Pleistocene temperature depression and vegetation change in Ecuadorian Amazonia. Quaternary Research 34:330–345.

Cabrera, A.
1961 Catálogo de los mamíferos de América del Sur. Revista del Museo Argentino de Ciencias Naturales "Bernardino Rivadavia", Ciencias Zoológicas 4 (2):309–732.

Carvalho, G. A. S.
1999 Relações filogenéticas entre formas recentes e fósseis de Echimyidae (Rodentia: Hystricognathi) e aspectos da evolução da morfologia dentária. Masters Thesis, Universidade Federal do Rio de Janeiro, Rio de Janeiro. 299 pp.

Clapperton, C. M.
1993 Quaternary Geology and Geomorphology of South America. Elsevier Science Publishers, Amsterdam. 779 pp.

Coimbra-Filho, A. F., and I. G. Câmara
1996 Os limites originais do bioma Mata Atlântica no Nordeste brasileiro. Fundação Brasileira para a Conservação da Natureza, Rio de Janeiro. 86 pp.

Colinvaux, P. A., P. E. Oliveira, J. E. Moreno, M. C. Miller, and M. B. Bush
1996a A long pollen record from lowland Amazonia: forest and cooling in glacial times. Science 275:85–88.

Colinvaux, P. A., K. B. Liu, P. E. Oliveira, M. B. Bush, M. C. Miller, and M. Steinitz Kannan
1996b Temperature depression in the lowland tropics in glacial times. Climatic Change 32:19–33.

Colinvaux, P. A., P. E. Oliveira, and M. B. Bush
  2000 Amazonian and Neotropical plant communities on glacial time-scales: the failure of the aridity and refuge hypothesis. Quaternary Science Reviews 19:141–169.

Colinvaux, P. A., and P. E. Oliveira
  2001 Amazon plant diversity and climate through the Cenozoic. Palaeogeography, Palaeoclimatology, Palaeoecology 166:51–63.

Conservation International do Brasil, Fundação SOS Mata Atlântica, Fundação Biodiversitas, Instituto de Pesquisas Ecológicas, Secretaria do Meio Ambiente–SP, and Instituto Estadual de Florestas–MG
  2000 Avaliação e ações prioritárias para a conservação da biodiversidade da Mata Atlântica e Campos Sulinos. Ministério do Meio Ambiente, Brasília. 40 pp.

Corbet, G. B.
  1997 The species in mammals. Pp. 341–356 in Species: The Units of Biodiversity (M. F. Claridge, H. A. Dawah, and M. R. Wilson, eds.). Chapman and Hall, London.

Costa, L. P., Y. L. R. Leite, G. A. B. Fonseca, and M. T. Fonseca
  2000 Biogeography of South American forest mammals: endemism and diversity in the Atlantic forest. Biotropica 32:872–881.

Cracraft, J.
  1983 Species concepts and speciation analysis. Pp. 159–187 in Current Ornithology. Plenum Press, New York.
  1989 Speciation and its ontology: the empirical consequences of alternative species concepts for understanding patterns and processes of differentiation. Pp. 28–59 in Speciation and Its Consequences (D. Otte and J. A. Endler, eds.). Sinauer Associates, Sunderland.

Cuvier, G. F.
  1809 Extrait des premiers Mémoires de M. F. Cuvier, sur les dents des mammifères consideerés comme charactères génériques. Noveau Bulletin des Sciences, par la Société Philomathique 1(24): 394-395.
  1838 Rapport sur un mémoir de M. Jourdan, de Lyon, concernant quelques mammifères nouveaux. Annales des Sciences Naturelles, ser. 2, 8:367–374.

Davis, D. E.
  1945 The annual cycle of plants, mosquitoes, birds, and mammals in two Brazilian forests. Ecological Monographs 15:243–295.

Dean, W.
  1995 With Broadax and Firebrand: The Destruction of the Brazilian Atlantic Forest. University of California Press, Berkeley. 482 pp.

Desmarest, A. G.
  1817 Nouveau dictionnaire d'histoire naturelle, appliquée aux arts, à l'agriculture, à l'économie rurale et domestique, à la médecine, etc. Chez Deterville, Paris.

Dietz, J. M., C. A. Peres, and L. Pinder
  1997 Foraging ecology and use of space in wild golden lion tamarins (*Leontopithecus rosalia*). American Journal of Primatology 41:289–305.

Eldredge, N., and J. Cracraft
  1980 Phylogenetic Analysis and the Evolutionary Process. Columbia University Press, New York. 349 pp.

Ellerman, J. R.
  1940 The Families and Genera of Lving Rodents. Volume 1: Rodents Other than Muridae. British Museum (Natural History), London. 689 pp.

Emmons, L. H.
  1981 Morphological, ecological, and behavioral adaptations for arboreal browsing in *Dactylomys dactylinus* (Rodentia: Echimyidae). Journal of Mammalogy 62:183–189.
  in review A revision of the genera of arboreal Echimyidae (Rodentia: Echimyidae: Echimyinae); with descriptions of two new genera. In: Mammalian Diversification: From Population Genetics to Phylogeography (E. A. Lacey and P. Myers, eds.). University of California Press, Berkeley.

Emmons, L. H., and F. Feer
  1990 Neotropical Rainforest Mammals: A Field Guide. University of Chicago Press, Chicago. 279 pp.
  1997 Neotropical Rainforest Mammals: A Field Guide. 2d edition. University of Chicago Press, Chicago. 307 pp.

Emmons, L. H., Y. L. R. Leite, D. Kock, and L. P. Costa
  2002 A review of the named forms of *Phyllomys* (Rodentia: Echimyidae) with the description of a new species from coastal Brazil. American Museum Novitates 3380:1–40.

Endler, J. A.
  1977 Geographic Variation, Speciation, and Clines. Princeton University Press, Princeton.

Eriksson, T.
  1998 Autodecay, ver. 4.0. Department of Botany, Stockholm University, Stockholm.

Excoffier, L., and P. Smouse
  1994 Using allele frequencies and geographic subdivision to reconstruct gene genealogies within a species: molecular variance parsimony. Genetics 136:343–359.

Felsenstein, J.
  1985 Confidence limits on phylogenies: an approach utilizing the bootstrap. Evolution 39:783–791.

Fonseca, G. A. B.
  1985 The vanishing Brazilian Atlantic forest. Biological Conservation 34:17–34.

Fonseca, G. A. B., and M. C. M. Kierulff
  1989 Biology and natural history of Brazilian Atlantic forest small mammals. Bulletin of the Florida State Museum, Biological Sciences 34:99–152.

Fonseca, G. A. B., G. Hermman, and Y. L. R. Leite
  1999 Macrogeography of Brazilian mammals. Pp. 549–563 in Mammals of the Neotropics: The Central Neotropics (J. F. Eisenberg and K. H. Redford, eds.). University of Chicago Press, Chicago.

Fonseca, M. T., and E. L. Sábato
  1998 *Echimys (Nelomys) braziliensis* (Waterhouse, 1848). Pp. 166–167 in Livro vermelho das espécies ameaçadas de extinção do estado de Minas Gerais (A. B. M. Machado, G. A. B. Fonseca, R. B. Machado, L. M. S. Aguiar, and L. V. Lins, eds.). Fundação Biodiversitas, Belo Horizonte.

Freitas, C. A.
 1957 Notícia sobre a peste no Nordeste. Revista Brasileira de Malariologia e Doenças Tropicais 9:123–133.

Geoffroy Saint-Hilaire, I.
 1840 Notice sur les rongeurs épineux désignés par les auteurs sous les noms d'*Echimys*, *Loncheres*, *Heteromys* et *Nelomys*. Magasin de Zoologie 2:1–57.

Haffer, J.
 1969 Speciation in Amazonian forest birds. Science 165:131–137.
 1993 Time's cycle and time's arrow in the history of Amazonia. Biogeographica 69:15–45.

Hartwig, W. C., and C. Cartelle
 1996 A complete skeleton of the giant South American primate *Propithecus*. Nature 381:307–311.

Hennig, W.
 1966 Phylogenetic Systematics. University of Illinois Press, Urbana. 263 pp.

Hensel, R.
 1872 Beiträge zur Kenntniss der Säugethiere Süd-Brasiliens. Abhandlungen der Königl. Akademie der Wissenschaften, Berlin 1872(1873):1–130.

Hilton-Taylor, C.
 2000 2000 IUCN Red List of Threatened Species. The World Conservation Union/Species Survival Comission (IUCN/SSC), Gland. 61 pp.

Hueck, K.
 1972 As florestas da América do Sul: ecologia, composição e importância. Editora da Universidade de Brasília, Brasília. 466 pp.

Ihering, H.
 1897 A Ilha de São Sebastião. Revista do Museu Paulista 2:9–171.
 1898 Bibliographia (História natural e Anthropologia). Revista do Museu Paulista 3:505–507.

IUCN
 2001 IUCN Red List Categories and Criteria: Version 3.1. IUCN Species Survival Commission. IUCN, Gland. 30 pp.

Jourdan, C.
  1837 Mémoire sur quelques mammifères nouveaux. Comptes Rendus Hebdomadaires des Sèances de l'Academie des Sciences 15:521–524.

Kimura, M.
  1980 A single method for estimating evolutionay rate of base substitutions through comparative studies of nucleotide sequences. Journal of Molecular Evolution 16:111–120.

Laemmert, H. W., L. C. Ferreira, and R. M. Taylor
  1946 An epidemiological study of jungle yellow fever in an endemic area in Brazil. Part 2: Investigation of vertebrate hosts and arthropod vectors. American Journal of Tropical Medicine 26 (suppl.):23–69.

Lara, M. C., M. A. Bogan, and R. Cerqueira
  1992 Sex and age components of variation in *Proechimys cuvieri* (Rodentia: Echimyidae) from northern Brazil. Proceedings of the Biological Society of Washington 105:882–893.

Lara, M. C., J. L. Patton, and M. N. F. Silva
  1996 The simultaneous diversification of South American echimyid rodents (Hystricognathi) based on complete cytochrome b sequences. Molecular Phylogenetics and Evolution 5:403–413.

Lara, M. C., and J. L. Patton
  2000 Evolutionary diversification of spiny rats (genus *Trinomys*, Rodentia: Echimyidae) in the Atlantic Forest of Brazil. Zoological Journal of the Linnean Society 130:661–686.

Leite, Y. L. R., L. P. Costa, and J. R. Stallings
  1996 Diet and vertical space use of three sympatric opossums in a Brazilian Atlantic forest reserve. Journal of Tropical Ecology 12:435–440.

Leite, Y. L. R., and J. L. Patton
  2002. Evolution of South American spiny rats (Rodentia, Echimyidae): the star-phylogeny hypothesis revisited. Molecular Phylogenetics and Evolution 25: 455–464.

Lund, P. W.
  1839a Coup-d'oeil sur le espèces éteintes de mammifères du Brésil; extrait de quelques mémoires présentés à l'Académie royale des Sciences de Copenhague. Annales des Sciences Naturelles, ser. 2 11:214–234.

1839b Nouvelles observations sur la faune fossile du Brésil; extraits d'une lettre adressée aux rédacteurs par M. Lund. Annales des Sciences Naturelles, ser. 2, 12:205–208.

1840a Blik paa Brasiliens Dyreverden för sidste Jordomvaeltning. Tredie Afhandling: Fortsaettelse af Pattedyrene. Det Kongelige Danske Videnskabernes Selskabs Naturvidenskabelige og Mathematiske Afhandlinger 8:217–272.

1840b Tillaeg til de to Sidste Afhandlinger over Brasiliens Dryeverden för sidste Jordomvaeltning. Det Konigelige Danske Videnskabernes Selskabs Naturvidenskabelige og Mathematiske Afhandlinger 8:273–296.

1840c Nouvelles recherches sur la faune fossile du Brésil. Annales des Sciences Naturelles, ser. 2, 13:310–319.

Lynch, J. D.
1988 Refugia. Pp. 311–342 in Analytical Biogeography: An Integrated Approach to the Study of Animal and Plant Distributions (A. A. Myers and P. S. Giller, eds.). Chapman and Hall, London.

1989 The gauge of speciation: on the frequencies of modes of speciation. Pp. 527–553 in Speciation and Its Consequences (D. Otte and J. A. Endler, eds.). Sinauer Associates, Sunderland.

Mayden, R. L.
1997 A hierarchy of species concepts: the denoument in the saga of the species problem. Pp. 381–424 in Species: The Units of Biodiversity (M. F. Claridge, H. A. Dawah and M. R. Wilson, eds.). Chapman and Hall, London.

Mayr, E.
1942 Systematics and the Origin of the Species from the Viewpoint of a Zoologist. Columbia University Press, New York. 334 pp.

McKenna, M. C., and S. K. Bell
1997 Classification of Mammals above the Species Level. Columbia University Press, New York. 631 pp.

Mishler, B. D., and M. J. Donoghue
1982 Species concepts: a case for pluralism. Systematic Zoology 31:491–503.

Mishler, B. D., and E. C. Theriot
2000 The Phylogenetic Species Concept (*sensu* Mishler and Theriot): monophyly, apomorphy, and phylogenetic species concepts. Pp. 44–54 in

Species Concepts and Phylogenetic Theory: A Debate (Q. D. Wheeler and R. Meier, eds.). Columbia University Press, New York.

Mittermeier, R. A., N. Myers, J. B. Thomsen, G. A. B. Fonseca, and S. Olivieri
 1998 Biodiversity hotspots and major tropical wilderness areas: approaches to setting conservation priorities. Conservation Biology 12:516–520.

Moffett, M. W.
 2000 "What's up"? A critical look at the basic terms of canopy biology. Biotropica 32:569–596.

Moojen, J.
 1950 "*Echimys* (*Phyllomys*) *kerri*" n. sp. (Echimyidae, Rodenia). Revista Brasileira de Biologia 10:489–492.
 1952 Os roedores do Brasil. Instituto Nacional do Livro, Rio de Janeiro. 214 pp.

Morellato, L. P. C., and C. F. B. Haddad
 2000 Introduction: the Brazilian Atlantic Forest. Biotropica 32:786–792.

Mori, S. A.
 1989 Eastern, extra-Amazonian Brazil. Pp. 428–454 in Floristic Inventory of Tropical Countries: The Status of Plant Systematics, Collections, and Vegetation, plus Recommendations for the Future (D. G. Campbell and H. D. Hammond, eds.). New York Botanical Garden, New York.

Mori, S. A., B. M. Boom, and G. T. Prance
 1981 Distribution patterns and conservation of eastern Brazilian coastal forest tree species. Brittonia 33:233–245.

Musser, G. G., M. D. Carleton, E. M. Brothers, and A. L. Gardner
 1998 Systematic studies of oryzomyine rodents (Muridae, Sigmodontinae): diagnoses and distributions of species formerly assigned to *Oryzomys* "*capito*." Bulletin of the American Museum of Natural History 236:1–376.

Mustrangi, M. A., and J. L. Patton
 1997 Phylogeography and systematics of the slender mouse opossum *Marmosops* (Marsupialia, Didelphidae). University of California Publications in Zoology 130:1–86.

Myers, N., R. A. Mittermeier, C. G. Mittermeier, G. A. B. Fonseca, and J. Kent
  2000 Biodiversity hotspots for conservation priorities. Nature 403:853–858.

Myers, P.
  1982 Origins and affinities of the mammal fauna of Paraguay. Pp. 85–93 in Mammalian Biology in South America (M. A. Mares and H. H. Genoways, eds.). University of Pittsburgh, Pittsburgh.

NIMA
  1997 Digital Interim Geographic Names Data (CDROM). U.S. Board on Geographic Names, U.S. National Imagery and Mapping Agency, Bethesda.

Nowak, R. M.
  1991 Walker's Mammals of the World, 5th ed., vol. 2. Johns Hopkins University Press, Baltimore.

Oliveira, T. G., and E. R. L. Mesquita
  1998 Notes on the distribution of the white-faced tree rat, *Echimys chrysurus* (Rodentia, Echimyidae) in northeastern Brazil. Mammalia 62:305–306.

Olmos, F.
  1996 Missing species in São Sebastião Island, southeastern Brazil. Papéis Avulsos de Zoologia 39:329–349.
  1997 The giant Atlantic forest tree rat *Nelomys thomasi* (Echimyidae): a Brazilian insular endemic. Mammalia 61:130–134.

Parizzi, M. G., M. L. Salgado-Labouriau, and H. C. Kohler
  1998 Genesis and environmental history of Lagoa Santa, southeastern Brazil. The Holocene 8:311–321.

Patton, J. L.
  1967 Chromosome studies of certain pocket-mice, genus *Perognathus* (Rodentia: Heteromyidae). Journal of Mammalogy 48:27–37.

Patton, J. L., and M. A. Rogers
  1983 Systematic implications of non-geographic variation in the spiny rat genus *Proechimys*. Zeitschrift fuer Saeugetierkunde 48:363–370.

Patton, J. L., and M. N. F. Silva
  1997 Definition of species of pouched four-eyed opossums (Didelphidae, *Philander*). Journal of Mammalogy 78:90–102.

Patton, J. L., M. N. F. Silva, M. C. Lara, and M. A. Mustrangi
  1997 Diversity, differentiation, and the historical biogeography of nonvolant small mammals of the neotropical forests. Pp. 455–465 in Tropical Forest Remnants: Ecology, Management, and Conservation of Fragmented Communities (W. F. Laurance and R. O. Bierregaard, eds.). University of Chicago Press, Chicago.

Patton, J. L., M. N. F. Silva, and J. R. Malcolm
  2000 Mammals of the Rio Juruá and the evolutionary and ecological diversification of Amazonia. Bulletin of the American Museum of Natural History 244:1–306.

Paula Couto, C.
  1950 Peter Wilhelm Lund: Memórias sobre a Paleontologia Brasileira. Instituto Nacional do Livro, Rio de Janeiro, Brazil. 591 pp.

Paynter, R. A., Jr., and M. A. Traylor, Jr.
  1991 Ornithological Gazetteer of Brazil. Museum of Comparative Zoology, Harvard University, Cambridge. 789 pp.

Pennington, R. T., D. E. Prado, and C. A. Pendry
  2000 Neotropical seasonally dry tropical forests and Quaternary vegetation changes. Journal of Biogeography 27:261–273.

Pessoa, L. M., and S. F. Reis
  1991 The contribution of cranial indedterminant growth to non-geographic variation in adult *Proechimys albispinus* (Is. Geoffroy) (Rodentia: Echimyidae). Zeitschrift für Saeugetierkunde 56:219–224.

Petri, S., and V. J. Fúlfaro
  1983 Geologia do Brasil (Fanerozóico). Editora da Universidade de São Paulo, São Paulo. 631 pp.

Por, F. D.
  1992 Sooretama, the Atlantic Rain Forest of Brazil. SPB Academic Publishing, The Hague. 130 pp.

Posada, D., and K. A. Crandall
  1998 Modeltest: testing the model of DNA substitution. Bioinformatics 14:817–818.

Prado, D. E., and P. E. Gibbs
  1993 Patterns of species distributions in the dry seasonal forests of South
        America. Annals of the Missouri Botanical Garden 80:902–927.

Prance, G. T. (ed.)
  1982 Biological Diversification in the Tropics. Columbia University Press,
        New York. 714 pp.

Purvis, A., and A. Hector
  2000 Getting the measure of biodiversity. Nature 405:212–219.

Queiroz, K., and M. J. Donoghue
  1988 Phylogenetic systematics and the species problem. Cladistics 4:317–338.

Reinhardt, J.
  1849 Iagttagelser om en besynderlig hyppig, abnorm Haleløshed hos flere
        brasilianske Pigrotter. Vidensk. Meddel. Naturhist. Foren. Kbhvn.:110–
        115.

Reiseberg, L. H.
  2001 Chromosomal rearrangements and speciation. Trends in Ecology and
        Evolution 16:351–358.

Rodríguez, F., J. F. Oliver, A. Marín, and J. R. Medina
  1990 The general stochastic model of nucleotide substitution. Journal of
        Theoretical Biology 142:485–501.

Rogers, A. R.
  1995 Genetic evidence for a Pleistocene population explosion. Evolution
        49:608–615.

Rogers, A. R., and H. C. Harpending
  1992 Population growth makes waves in the distribution of pairwise genetic
        differences. Molecular Biology and Evolution 9:552–569.

Rosen, D.
  1978 Vicariant patterns and historical explanation in biogeography. Systematic
        Zoology 27:159–188.

Safford, H. D.
　1999a Brazilian Páramos I. An introduction to the physical environment and vegetation of the campos de altitude. Journal of Biogeography 26:693–712.
　1999b Brazilian Páramos II. Macro- and mesoclimate of the campos de altitude and affinities with high mountain climates of the tropical Andes and Costa Rica. Journal of Biogeography 26:713–737.

Salgado-Labouriau, M. L., M. Barberi, K. R. Ferraz-Vicentini, and M. G. Parizzi
　1998 A dry climatic event during the late Quaternary of tropical Brazil. Review of Palaeobotany and Palynology 99:115–129.

Sbalqueiro, I. J., A. M. S. Bueno, J. Moreira, A. P. D. Ramos, C. Padovani, A. Ximenez, and J. M. S. Agostini
　1988 Cariótipo com 96 cromossomos em *Echimys dasythrix*, o mais elevado número diplóide entre os mamíferos. Resumos do XV Congresso Brasileiro de Zoologia, Curitiba, p. 532.

Schneider, S., D. Roessli, and L. Excoffier
　2000 Arlequin 2.000. Department of Anthropology and Ecology, University of Geneva, Geneva.

Silva, L. F. B. M.
　1993 Ecologia do rato do bambu, *Kannabateomys amblyonyx* (Wagner, 1845), na Reserva Biológica de Poço das Antas, Rio de Janeiro. Masters Thesis, Universidade Federal de Minas Gerais, Belo Horizonte. 80 pp.

Silva, M. N. F.
　1998 Four new species of spiny rats of the genus *Proechimys* (Rodentia: Echimyidae) from the western Amazon of Brazil. Proceedings of the Biological Society of Washington 111:436–471.

Silva, M. N. F., and J. L. Patton
　1998 Molecular phylogeography and the evolution and conservation of Amazonian mammals. Molecular Ecology 7:475–486.

Souza, M. J.
　1981 Caracterização cromossômica em oito espécies de roedores brasileiros das famílias Cricetidae e Echimyidae. Ph.D. Dissertation, Universidade de São Paulo, São Paulo.

Swofford, D. L.
  2000 PAUP*. Phylogenetic Analysis Using Parsimony (*and Other Methods).
        version 4.0. Sinauer Associates, Sunderland.

Tajima, F.
  1989 Statistical method for testing the neutral mutation hypothesis by DNA
        polymorphisms. Genetics 123:585–595.

Tate, G. H. H.
  1935 The taxonomy of the genera of Neotropical hystricoid rodents. Bulletin
        of the American Museum of Natural History 68:295–447.

Thomas, O.
  1909 Notes on some South American mammals, with descriptions of new
        species. Annals and Magazine of Natural History, ser. 8, 4:230–242.
  1916a Some notes on the Echimyinae. Annals and Magazine of Natural History,
        ser. 8, 18:294–301.
  1916b On the generic names *Rattus* and *Phyllomys*. Annals and Magazine of
        Natural History, ser. 8, 18:240.

Vane-Wright, R. I., C. J. Humphries, and P. H. Williams
  1991 What to protect? Systematics and the agony of choice. Biological
        Conservation 55:235–254.

Vanzolini, P. E.
  1970 Zoologia sistemática, geografia e a origem das espécies. Teses e
        Monografias, Instituto Geográfico de São Paulo 3:1–56.
  1988 Distributional patterns of South American lizards. Pp. 317–342 in
        Proceedings of a Workshop on Neotropical Distribution Patterns (P. E.
        Vanzolini and W. R. Heyer, eds.). Academia Brasileira de Ciências, Rio
        de Janeiro.
  1992 A Supplement to the Ornithological Gazetteer of Brazil. Museu de
        Zoologia, Universidade de São Paulo, São Paulo,

Vieira, C. C.
  1955 Lista remissiva dos mamíferos do Brasil. Arquivos de Zoologia do
        Estado de São Paulo 8:341–474.

Vivo, M.
  1996 How many species of mammals are there in Brazil? Taxonomic practice
        and diversity evaluation. Pp. 313–321 in Biodiversity in Brazil: A First
        Approach (C. E. M. Bicudo and N. A. Menezes, eds.). CNPq, São Paulo.

1997   Mammalian evidence of historical ecological change in the Caatinga semiarid vegetation of northeastern Brazil. Journal of Comparative Biology 2:65–73.

Voss, R. S.
1993   A revision of the Brazilian muroid rodent genus *Delomys* with remarks on thomasomyine characters. American Museum Novitates 3073:1–44.

Voss, R. S., and L. H. Emmons
1996   Mammalian diversity in Neotropical lowland rainforests: a preliminary assessment. Bulletin of the American Museum of Natural History 230:1–115.

Voss, R. S., D. P. Lunde, and N. B. Simmons
2001   The mammals of Paracou, French Guiana: a Neotropical lowland rainforest fauna. Part 2. nonvolant species. Bulletin of the American Museum of Natural History 263:1–236.

Wagner, A.
1842   Diagnosen neuer Arten brasilischer Säugthiere. Archiv für Naturgeschichte 1:356–362.

Wheeler, Q. D., and R. Meier (eds.)
2000   Species Concepts and Phylogenetic Theory: A Debate. Columbia University Press, New York.

White, M. J. D.
1968   Models of speciation. Science 159:1065–1070.

Whitmore, T. C., and G. T. Prance (eds.)
1987   Biogeography and Quaternary History in Tropical America. Clarendon Press, Oxford. 214 pp.

Winge, H.
1887   Jordfundne og nulevende gnavere (Rodentia) fra Lagoa Santa, Minas Geraes, Brasilien: med udsigt over gnavernes indbyrdes slægtskab. E Museo Lundii 1(3):200 pp.

Woods, C. A.
1993   Suborder Hystricognathi. Pp. 771–806 in Mammal Species of the World: A Taxonomic and Geographic Reference (D. E. Wilson and D. M. Reeder, eds.). Smithsonian Institution Press, Washington, D.C.

Yonenaga, Y.
  1975 Karyotypes and chromosome polymorphism in Brazilian rodents. Caryologia 28:269–286.

Zanchin, N. I.
  1988 Estudos cromossômicos em orizominos e equimídeos da Mata Atlântica. Master's Thesis, Universidade Federal do Rio Grande do Sul, Porto Alegre.

Zimmermann, E. A. W.
  1780 Geographische Geschichte des Menschen, und der allgemein verbreiteten vierfüssigen Thiere: nebst einer hieher gehörigen zoologischen Weltcharte. In der Weygandschen Buchhandlung, Leipzig.

DATE DUE

GAYLORD   PRINTED IN U.S.A.